REVISE AQA GCSE (9–1)
Chemistry

REVISION WORKBOOK

Higher

Series Consultant: Harry Smith

Authors: Nora Henry

Also available to support your revision:

Revise GCSE Study Skills Guide　　　9781447967071

The **Revise GCSE Study Skills Guide** is full of tried-and-trusted hints and tips for how to learn more effectively. It gives you techniques to help you achieve your best – throughout your GCSE studies and beyond!

Revise GCSE Revision Planner　　　9781447967828

The **Revise GCSE Revision Planner** helps you to plan and organise your time, step-by-step, throughout your GCSE revision. Use this book and wall chart to mastermind your revision.

For the full range of Pearson revision titles across KS2, KS3, GCSE, Functional Skills, AS/A Level and BTEC visit: www.pearsonschools.co.uk/revise

Contents

A small bit of small print:

AQA publishes Sample Assessment Material and the Specification on its website. This is the official content and this book should be used in conjunction with it. The questions in this book have been written to help you practise every topic in the book. Remember: the real exam questions may not look like this.

Elements, mixtures and compounds

1 What is a compound?

Tick **one** box.

☐ a substance made up of only one element

☐ a substance made up of two or more elements chemically joined together

☐ a substance made up of only metallic elements chemically joined together

☐ a substance made up of two or more elements physically mixed in a fixed ratio   **(1 mark)**

> Question 1 is an example of a multiple-choice question, in which you only need to tick the correct answer. There will be many multiple-choice questions on both of your papers.

2 Some symbols and formulae are given in the box below. Write a formula or symbol from the box for:

NaOH	Al_2O_3	H_2O	Na	S	NH_3	CO

(a) a metallic element...   **(1 mark)**

(b) a molecule containing four atoms..   **(1 mark)**

(c) a compound containing three different elements. ....   **(1 mark)**

3 When a mixture of the elements iron and sulfur is heated, a compound is formed.

(a) Name the compound formed. ....   **(1 mark)**

> Guided

(b) The element iron is attracted to magnets. The element sulfur is not affected by magnets. When a magnet is pulled through a mixture of powdered iron and sulfur, clumps of dark powder stick to the magnet and the powder left behind has a yellow colour. When a magnet is pulled through the powdered compound described in question 3(a), no such colour separation is observed. Explain these observations.

In the mixture, the iron powder will cluster on the magnet, while the sulfur will

remain behind. However, compounds can only be separated by chemical

reactions, so ....

....   **(3 marks)**

4 The diagram represents particles present in four different substances. Circles represent atoms and different shades of grey represent different elements.

Which box, A, B, C or D, represents:

(a) particles of a mixture of two different elements? ....   **(1 mark)**

(b) molecules of the compound NO? ....

> The compound NO contains two different types of atoms, which are chemically joined.

   (1 mark)

(c) molecules of a mixture of two compounds? ....   **(1 mark)**

1

Practical skills

Filtration, crystallisation and chromatography

1 The diagram shows some different apparatus used to separate mixtures.

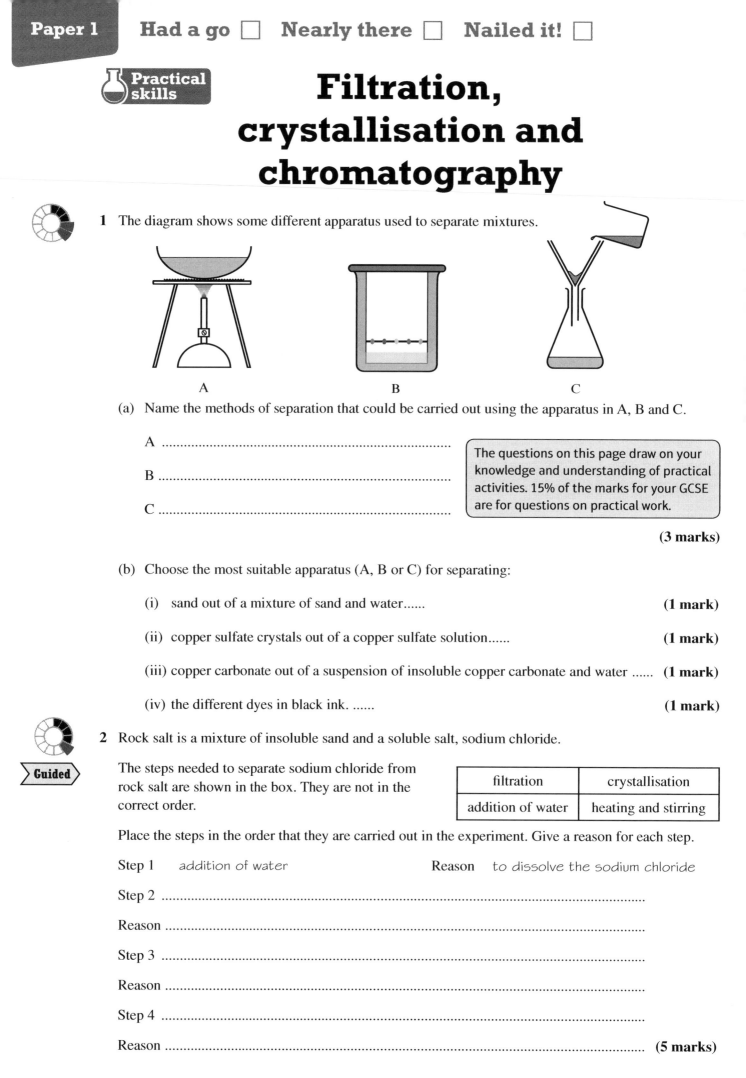

A   B   C

(a) Name the methods of separation that could be carried out using the apparatus in A, B and C.

A ....

B ....

C ....

> The questions on this page draw on your knowledge and understanding of practical activities. 15% of the marks for your GCSE are for questions on practical work.

(3 marks)

(b) Choose the most suitable apparatus (A, B or C) for separating:

(i) sand out of a mixture of sand and water...... **(1 mark)**

(ii) copper sulfate crystals out of a copper sulfate solution...... **(1 mark)**

(iii) copper carbonate out of a suspension of insoluble copper carbonate and water ...... **(1 mark)**

(iv) the different dyes in black ink. ...... **(1 mark)**

2 Rock salt is a mixture of insoluble sand and a soluble salt, sodium chloride.

Guided

The steps needed to separate sodium chloride from rock salt are shown in the box. They are not in the correct order.

filtration	crystallisation
addition of water	heating and stirring

Place the steps in the order that they are carried out in the experiment. Give a reason for each step.

Step 1   addition of water   Reason   to dissolve the sodium chloride

Step 2 ....

Reason ....

Step 3 ....

Reason ....

Step 4 ....

Reason .... **(5 marks)**

 **Practical skills**

Distillation

1 Two separation techniques are shown in the diagram below. The diagrams are not labelled.

(a) What is the name of the technique carried out using the apparatus on the right of the diagram?

Tick **one** box.

☐ chromatography   ☐ distillation

☐ crystallisation   ☐ fractional distillation   **(1 mark)**

(b) What is the purpose of the piece of apparatus labelled A?

.... **(1 mark)**

(c) What change of state happens at B?

.... **(1 mark)**

(d) No labels have been included in the diagram above.
Name the labels that should be placed at:

> **Practical skills** It is important that you know how to label diagrams for all methods of separation.

C .... **(1 mark)**

D .... **(1 mark)**

E .... **(1 mark)**

(e) Name a different way of heating the apparatus shown in the diagram above.

.... **(1 mark)**

2 Distillation can be used to obtain water from salty water. Describe how the technique of distillation works.

> Refer to changes of state in this question.

> Guided

The salty water is heated and the water...

....

.... **(3 marks)**

Historical models of the atom

1   The 'plum pudding' model of an atom, shown in the diagram, suggested that the atom was a charged ball.

charged ball —

A

(a)   What type of charge was thought to be on the ball in the plum pudding model?

.... **(1 mark)**

(b)   Name particle A.

.... **(1 mark)**

2   Rutherford and Marsden carried out an experiment by firing positively charged alpha particles at a thin metal foil. They observed that some alpha particles passed straight through the foil, others bounced back and others changed direction. Describe the deductions they were able to make about the atom from these observations.

....
....

> Remember that like charges repel.

....

.... **(3 marks)**

3   The neutron was the last subatomic particle to be discovered.

(a)   Name the scientist who discovered evidence for the existence of the neutron.

.... **(1 mark)**

(b)   Name **one** feature of a neutron that made it difficult to detect and suggest why.

.... **(2 marks)**

4   Describe the difference between the plum pudding model of the atom and the nuclear model of the atom.

> Guided

The plum pudding model suggested that an atom was a ball of ....

The nuclear model has positive protons in a ....

....

....

> You now need to mention the position of electrons in the plum pudding model and how this differs from their position in the nuclear model.

....

.... **(4 marks)**

Particles in an atom

1 The element sodium has an atomic number of 11 and a mass number of 23.

(a) Define mass number.

.... **(1 mark)**

(b) In terms of subatomic particles, why does a sodium atom have no overall charge?

.... **(1 mark)**

(c) Give the number of protons, neutrons and electrons in a sodium atom.

> Remember, for a neutral atom, the number of protons equals the number of electrons.

> To find the number of neutrons subtract the atomic number from the mass number.

Number of protons ....

Number of neutrons ....

Number of electrons .... **(3 marks)**

(d) Name **two** subatomic particles found in the nucleus of a sodium atom.

.... **(1 mark)**

2 (a) Complete the table below to give the number of protons, neutrons and electrons in each of four different atoms, A, B, C and D.

> **Guided**

Atom	Atomic number	Mass number	Number of electrons	Number of neutrons	Number of protons
A	27	59	27	59 − 27 = 32	27
B	28	59			
C	13	27			
D	19	39			

(4 marks)

(b) Use the periodic table on page 122 to give the name of each atom.

A   cobalt

> The atomic number identifies an atom. For A the atomic number is 27, which is cobalt.

B   ....

C   ....

D   .... **(4 marks)**

3 The diagram opposite shows a lithium atom.

> **Guided**

Describe the structure of this atom, stating the names and number of each particle present.

nucleus

The lithium atom is made up of a central nucleus containing ....

and .... Around the nucleus there are .... **(3 marks)**

Atomic structure and isotopes

1 An atom of potassium has the symbol $^{39}_{19}$K.

   (a) Complete the table to show the relative mass and charge of each particle present in a potassium atom.

Particle	Relative mass	Relative charge
electron		
neutron		
proton		

(3 marks)

   (b) Give the number of protons, neutrons and electrons in this atom of potassium.

   Number of protons ....

   Number of neutrons ....

   Number of electrons .... **(3 marks)**

   (c) Give the approximate radius of a potassium atom.

   Give your answer in metres and nanometres.

   > **Maths skills** Remember that 1 nanometre = 1×10^{-9} m

   .... **(2 marks)**

   (d) Another atom of potassium has the symbol $^{41}_{19}$K. Explain why atoms of $^{41}_{19}$K and $^{39}_{19}$K are isotopes.

   .... **(2 marks)**

2 Carbon has two naturally occurring isotopes, ^{12}C and ^{13}C.

   (a) Why do these carbon isotopes react in the same way?

   .... **(1 mark)**

   (b) Describe the differences and similarities in the isotopes ^{12}C and ^{13}C. You should refer to the subatomic particles present.

   > You need to calculate the number of protons, neutrons and electrons in each isotope.

   .... **(2 marks)**

   > **Guided**

   (c) Use the information about the two isotopes of carbon in the table below to calculate the relative atomic mass of carbon to one decimal place.

Mass number	12	13
Abundance	99	1

   > **Maths skills** Remember, when rounding to one decimal place, if the second decimal place number is 5 or more, round up.

   Relative atomic mass =

   $$\frac{\text{mass number isotope 1} \times \text{abundance} + \text{mass number isotope 2} \times \text{abundance}}{\text{total abundance}}$$

   $$= \frac{12 \times 99 + \text{.............................}}{(99 + 1)} = \text{..........................}$$ **(2 marks)**

Electronic structure

1 Complete the energy level (shell) diagrams below for the elements with the following number of electrons:

Guided

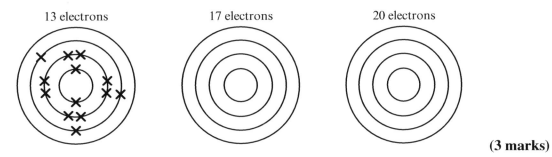

13 electrons   17 electrons   20 electrons

(3 marks)

2 Use the periodic table on page 122 to help you answer this question.

The diagram below shows the electron structure of an atom of an element.

> To find the number of neutrons, you need to get the mass number from the periodic table and subtract the atomic number from it.

(a) What is the name and atomic number of this element?

.... **(2 marks)**

(b) State the number of protons and electrons in the atoms of this element.

.... **(1 mark)**

(c) Calculate the number of neutrons in the atoms of this element.

.... **(1 mark)**

3 The electronic structure of magnesium can be written as 2,8,2. Write the electronic structures for the following elements in the same way.

(a) potassium .... **(1 mark)**

(b) phosphorus .... **(1 mark)**

4 Complete the missing information in the table below.

Element's name	Lithium	Aluminium	
Atomic number		13	
Diagram			
Electronic structure			2,8

(4 marks)

Development of the periodic table

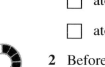

1   Mendeleev listed the elements in his periodic table in an order. Which property did he use to list the elements?

Tick **one** box.

☐ atomic number   ☐ mass number

☐ atomic weight   ☐ number of neutrons   **(1 mark)**

2   Before the discovery of protons, neutrons and electrons, scientists attempted to classify the elements by arranging them in order of their atomic weights.

(a)   Why did arranging elements in strict order of atomic weights not result in the development of an accurate periodic table?

....

.... **(2 marks)**

(b)   Mendeleev overcame some of these problems in creating his periodic table, right.

State **two** important steps that Mendeleev took when constructing his periodic table that helped overcome some of the problems with previous arrangements of the elements.

....

....

.... **(2 marks)**

	Group						
	1	2	3	4	5	6	7
	H						
	Li	Be	B	C	N	O	F
	Na	Mg	Al	Si	P	S	Cl
	K	Ca	*	Ti	V	Cr	Mn
	Cu	Zn	*	*	As	Se	Br
	Rb	Sr	Y	Zr	Nb	Mo	*
	Ag	Cd	In	Sn	Sb	Te	I

(c)   Suggest a reason why zinc was grouped with magnesium and calcium.

> First think about why magnesium and calcium are placed in the same group and apply this fact to zinc.

.... **(1 mark)**

(d)   In the periodic table above, the element in Group 4 marked by an asterisk (*) has the atomic number 32. Name this element.

.... **(1 mark)**

3   Compare Mendeleev's periodic table with the modern periodic table.

Guided

Mendeleev listed elements in his periodic table by ....

....

In Mendeleev's periodic table there is no group ....

....

In Mendeleev's periodic table there are gaps ....

....

.... **(4 marks)**

The modern periodic table

1 How are elements arranged in the modern periodic table?

Tick **one** box.

☐ by increasing atomic number

☐ by increasing mass number

☐ by increasing number of neutrons

☐ by increasing reactivity   **(1 mark)**

2 The diagram below shows the position of six different elements in the periodic table. The letters do not represent the symbols for the elements.

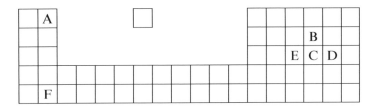

Use the letters in the diagram to answer the questions below.

(a) Identify **two** metals with the same number of electrons in the outer shell of their atoms.

> Remember that metals are found on the left of the periodic table.

.... **(1 mark)**

(b) Identify a halogen.

.... **(1 mark)**

(c) Identify a non-metal that has five electrons in the outer shell of its atoms.

.... **(1 mark)**

3 Group 3 of the periodic table contains the elements boron, aluminium and gallium.

(a) Why do these three elements have similar chemical properties?

....

.... **(1 mark)**

(b) Suggest why boron would have different physical properties from the other elements of Group 3.

> Look at the periodic table on page 122. Do you notice anything about B and Al?

.... **(1 mark)**

> Guided

(c) Which of boron, aluminium and gallium has the lowest number of protons in the nucleus?

The element with the smallest atomic number is .... As the

atomic number gives the number of protons, ... **(1 mark)**

Group 0

1 Which of the electronic structures below is the structure of an atom of a noble gas? Tick **one** box.

☐ 2   ☐ 2,2   ☐ 2,8,2   ☐ 2,8,7

(1 mark)

2 The table below shows some properties of the noble gases.

Element	Boiling point in °C	Density in g/dm³	Relative atomic mass
helium	−269	0.2	4
neon	−246	1.2	20
argon	−190		40
krypton		2.2	84
xenon	−111	5.9	131

(a) Predict the boiling point of krypton .... **(1 mark)**

(b) What is the relationship between boiling point and relative atomic mass?

.... **(1 mark)**

(c) Estimate the density of argon. .... **(1 mark)**

> Look at the general trend in density. It decreases down the group. You are asked for an estimate, so a value halfway between the values above and below is a good idea.

(d) Write the electronic structures of helium and neon.

.... **(2 marks)**

(e) Why is helium in the same group of the periodic table as neon and argon?

....

.... **(1 mark)**

3 Why do the atoms of noble gases not easily form molecules?

Tick **one** box.

> The first answer in this guided question has been crossed out. This is because it is untrue – helium is a noble gas and it has only two electrons in its outer shell.

☐ ~~They all have eight electrons in their outer shell.~~

☐ Their atoms have stable arrangements of electrons.

☐ The inner shells of electrons cannot be used in chemical bonding.

☐ Their reactivity decreases down the group.   **(1 mark)**

4 The diagram on the right shows the electronic structure of a noble gas atom.

(a) How many protons are in the nucleus of this atom? .... **(1 mark)**

(b) Use the periodic table on page 122 to identify this atom. .... **(1 mark)**

(c) How many neutrons are in the nucleus of this atom?

.... **(1 mark)**

> Use the periodic table on page 122 to find the mass number, then you can work out the number of neutrons.

(d) Why is this atom unreactive?...

.... **(1 mark)**

Group 1

1 Some alkali metals are shown in the box opposite. | lithium   potassium   rubidium   sodium |

    (a) Which of these metals is the most dense? .... **(1 mark)**

    (b) Which of these metals is the most reactive? .... **(1 mark)**

2 (a) Describe the bonding type, solubility and general colour of most compounds formed between alkali metals and non-metals.

....

....

.... **(3 marks)**

    (b) Describe and explain the trend in reactivity of the alkali metals (Group 1).

> It may be helpful to draw the electronic configurations of the first three alkali metals and compare the distance of the outer electron from the nucleus.

....

....

....

.... **(3 marks)**

3 Two elements in Group 1 of the periodic table are lithium and potassium.

    (a) Explain, in terms of their electronic configurations, why lithium and potassium are both in Group 1 of the periodic table.

.... **(1 mark)**

    (b) Very small pieces of lithium and potassium are separately allowed to react with water. Describe **two** similarities and **two** differences in what is observed, and name the products of the reactions.

> There are two things to do in this question: (1) compare the observations and (2) name the products.

....

....

....

.... **(4 marks)**

4 Sodium is a Group 1 metal that reacts with non-metals.

The word equation for the reaction of sodium with water is:

sodium + water → sodium hydroxide + hydrogen

Complete the balanced chemical equation for this reaction and name the ion that causes the final solution to be alkaline.

> Remember that hydrogen is diatomic. Then work out the formula of sodium hydroxide and balance the equation to make sure there is the same number of atoms on each side of the equation.

.... Na + ... H_2O → .... + H_2

Ion .... **(2 marks)**

> Guided

Group 7

1 Describe and explain the change in melting point down the Group 7 elements.

Guided

As you go down the group the melting point ....

This is because the forces between larger molecules are .... **(2 marks)**

2 An experiment involving the reactions of Group 7 elements with iron is shown in the diagram.

iron wool

a few crystals of iodine or a few drops of bromine

fumes released into fume cupboard

Group 7 element	Description of reaction with fine iron wool	Name and formula of main compound formed
fluorine	not carried out	–
chlorine	burns brightly	$FeCl_3$
bromine	glows red	$FeBr_2$
iodine	changes colour slowly	Iron(II) iodide, FeI_2

(a) (i) Why is the experiment carried out in a fume cupboard?

.... **(1 mark)**

(ii) Suggest **one** reason for not carrying out the reaction with fluorine.

.... **(1 mark)**

(b) Name the compounds of chlorine and bromine that are missing from the table.

.... **(2 marks)**

(c) Write a balanced chemical equation for the reaction of iron wool with iodine to form FeI_2.

.... **(2 marks)**

(d) Explain why the reactivity of the halogens decreases down the group.

....

....

.... **(2 marks)**

> Write down the electronic configuration of the first three halogens, and note what is happening to the outer shell. Do the halogens need to gain or lose electrons?

3 When chlorine is bubbled through a solution of sodium bromide, bromine is formed in solution.

(a) Complete the balanced symbol equation for this reaction.

$Cl_2(g) + 2NaBr(aq) \rightarrow$ .... (aq) + ... (aq)   **(2 marks)**

(b) What is this kind of reaction called? .... **(1 mark)**

(c) What is the ionic equation for the reaction of chlorine with sodium bromide?

Tick **one** box.

☐ $Cl_2 + 2Na \rightarrow 2NaCl$

☐ $2Br^- + Cl_2 \rightarrow Br_2 + 2Cl^-$

☐ $Br^- + Cl \rightarrow Br + Cl^-$

☐ $Br^- + Na^+ \rightarrow NaBr$ **(1 mark)**

Transition metals

1 Place a tick in either the true or false cell in the table for each of the following general statements, comparing the properties of transition metals with alkali metals.

Statements	True	False
Alkali metals have higher densities than transition metals.		
Transition metals are stronger than alkali metals.		
Alkali metals are harder than transition metals.		
Transition metals are more reactive than alkali metals.		

(4 marks)

2 This question refers to the elements in the different blocks of the periodic table shown below.

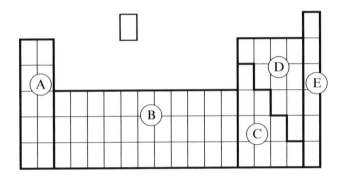

Which of the blocks A to E contain:

(a) transition metals .... **(1 mark)**

(b) the group of elements that usually form ions with a +1 charge .... **(1 mark)**

(c) metals that are often used as catalysts .... **(1 mark)**

(d) the most unreactive elements? .... **(1 mark)**

3 Copper(II) sulfate and sodium hydrogen carbonate are compounds.

> Guided

(a) Give the charge on the metallic ion in each compound and give the formula of each compound.

The metallic ion in copper(II) sulfate is copper, which has a 2+ charge. Sulfate has a charge of 2−, hence the formula of copper(II) sulfate is $CuSO_4$. In sodium hydrogen carbonate, the metallic ion is ....

.... **(2 marks)**

(b) Write down whether the two solids are coloured or white.

Copper(II) sulfate is .... and sodium hydrogen

carbonate is ....

> Remember, compounds containing a transition metal are almost always coloured.

(2 marks)

(c) Name **two** properties of (transition metal) copper that mean it can be used to make saucepans.

.... **(2 marks)**

Chemical equations

1 The chemical equation for the reaction between methane and oxygen is shown below.

$CH_4 + 2O_2 \rightarrow CO_2 + 2H_2O$

(a) Describe this reaction between methane and oxygen in terms of the names of the substances and the number of molecules involved.

.... **(2 marks)**

Guided

(b) When 4 g of methane burns, 11 g of carbon dioxide and 9 g of water are produced. What mass of oxygen was needed to react with the 4 g of methane?

> Remember, no atoms are gained or lost during a chemical reaction, so the total mass of reactants used up will always equal the total mass of products formed.

Mass of products   = 11 + 9
                = 20 g

Mass of reactants   = ... g, so mass of oxygen = ... g   **(1 mark)**

2 Ethane and propane both burn in air.

(a) Balance this symbol equation for the burning of ethane.

$C_2H_6 +$ .... $O_2 \rightarrow$ .... $CO_2 +$ .... H_2O **(2 marks)**

(b) Balance this symbol equation for the burning of propane.

$C_3H_8 +$ .... $O_2 \rightarrow$ .... $CO_2 +$ .... H_2O **(2 marks)**

(c) Name the products in these reactions.

.... **(2 marks)**

3 Sodium metal burns in oxygen to form sodium oxide:

.... $Na +$ .... $O_2 \rightarrow$ .... Na_2O

(a) Balance the above symbol equation for the reaction of sodium and oxygen.   **(2 marks)**

(b) Potassium burns in oxygen in a similar way to sodium. Write a balanced symbol equation for the reaction of potassium and oxygen.

.... **(2 marks)**

4 Balance the following equations.

(a) ...... $HCl +$ ...... $Ca \rightarrow$ ...... $CaCl_2 +$ ...... H_2 **(1 mark)**

(b) ...... $MgCO_3 +$ ...... $HCl \rightarrow$ ...... $MgCl_2 +$ ...... $H_2O +$ ...... CO_2 **(1 mark)**

(c) ...... $Al +$ ...... $HCl \rightarrow$ ...... $AlCl_3 +$ ...... H_2 **(1 mark)**

5 Write balanced symbol equations for the following reactions

(a) sodium + chlorine → sodium chloride .... **(2 marks)**

(b) potassium + water → potassium hydroxide + hydrogen .... **(2 marks)**

(c) magnesium + hydrochloric acid → magnesium chloride + hydrogen.......................... **(2 marks)**

Extended response – Atomic structure

Compare the chemical properties (including relative reactivity with water and charges on the ions) and the physical properties (including melting point, density and hardness) of the alkali metals with the transition metals.

> The question asks you to compare, so make sure you do this by considering properties that are similar and properties that are different. In your answers, use statements like 'The alkali metals …, but the transition metals…'.

....

....

....

....

....

....

....

....

....

....

....

....

....

....

....

....

....

....

....

....

.... **(6 marks)**

> Check your answer and make sure you have fully answered the question. It is a good idea to tick the parts of the question you have done, so you do not leave points out.

Forming bonds

1 Which pair of elements forms a covalent compound?

Tick **one** box.

    ☐ lithium and chlorine   ☐ carbon and oxygen

    ☐ magnesium and oxygen   ☐ potassium and bromine   **(1 mark)**

2 There are three types of strong chemical bonds, which are found in different substances. Name the type of strong chemical bonds found in:

(a) sodium chloride .... **(1 mark)**

(b) sodium .... **(1 mark)**

(c) carbon dioxide .... **(1 mark)**

(d) calcium chloride. .... **(1 mark)**

> Remember, metallic bonding occurs in metals (these have one-word names). Ionic bonding is between a metal and a non-metal, and covalent bonding occurs between two non-metals.

3 The diagram on the right shows the structure of a substance

(a) What type of bonding is represented in the diagram?

.... **(1 mark)**

(b) Identify particle A.

.... **(1 mark)**

(c) Identify particle B.

.... **(1 mark)**

4 Describe the difference between ionic bonding and covalent bonding.

Guided

Ionic bonding is the attraction between oppositely ....

Covalent bonding is the ....

.... **(2 marks)**

5 The bonding in a compound of nitrogen and hydrogen is shown in the diagram below.

(a) Write the formula of the compound shown in this diagram.

.... **(1 mark)**

(b) Name and describe the type of bonding shown.

....

.... **(1 mark)**

Ionic bonding

1 What happens when calcium reacts with chlorine to form calcium chloride?

Tick **one** box.

☐ Each chlorine atom loses one electron.   ☐ Each calcium atom loses one electron.

☐ Each chlorine atom gains one electron.   ☐ Each calcium atom gains two electrons.

(1 mark)

2 The table below shows some information for several atoms and simple ions. Complete the table using the periodic table on page 122 to help.

Atom/ion	Number of protons	Electronic configuration
	13	2,8,3
S²⁻		
Na⁺		
	12	2,8

Atoms and ions of the same element will have the same number of protons. When positive ions are formed, electrons are lost. When negative ions are formed, electrons are gained.

(3 marks)

3 Sodium forms an ionic compound with oxygen. Describe what happens when two atoms of sodium react with one atom of oxygen. Give the formulae of the ions formed.

Guided

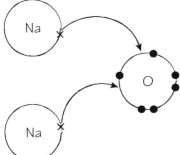

You need to work out the number of electrons in the outer shell, and think about their transfer, as shown in the diagram. Then, describe in words where the electrons transfer from and to, and how many electrons are involved.

Two sodium atoms each lose ....

....

....

.... **(5 marks)**

4 Some ions are listed in the box below.

aluminium ion	fluoride ion	magnesium ion	oxide ion	potassium ion

Choose **one** ion from the box to answer each question below. Each ion may be used once, more than once or not at all.

(a) Which ion has a charge of 2−? .... **(1 mark)**

(b) Which ion has the same number of electrons as an argon atom? .... **(1 mark)**

(c) Which ion has a charge of 3+? .... **(1 mark)**

(d) Which ion is formed from its atom by the loss of one electron? .... **(1 mark)**

Giant ionic lattices

1   What surrounds each sodium ion in a sodium chloride crystal?

Tick **one** box.

☐ one chloride ion   ☐ four chloride ions

☐ two chloride ions   ☐ six chloride ions   **(1 mark)**

2   The structure of caesium chloride can be represented
using the ball-and-stick model shown in the diagram on the right.

(a) What type of bonding is found in caesium chloride?

    ....   **(1 mark)**

(b) What is the name for this type of structure?

    ....   **(1 mark)**

(c) The ball-and-stick model is not a good representation of an ionic compound.

Give **one** reason why.

    ....

    ....   **(1 mark)**

3   A diagram of a solid sample of sodium chloride is
shown in the diagram on the right.

(a) Name **two** limitations of using this diagram to
represent a crystal of sodium chloride.

    ....

    .... **(2 marks)**

> What particles
> are found
> in sodium
> chloride? Does
> this diagram
> accurately show
> these? Think
> about their size
> and charge.

Guided   (b) Draw a dot-and-cross diagram to show the formation of sodium chloride
from atoms of sodium and chlorine. Only show outer shells in
your answer.

$$Na^{\bullet} \;+\; {}^{\times}_{\times}\!\overset{\times\times}{\underset{\times\times}{Cl}}{}^{\times}_{\times}$$

> The outer shell electrons are
> shown. Now show the transfer and
> the charge on each ion formed.

(4 marks)

(c) Explain why sodium chloride has a high melting point.

    ....

    ....

    ....   **(2 marks)**

Covalent bonding

1 A dot-and-cross diagram for the bonding in a molecule of the gas phosphine is shown below.

   (a) Complete the diagram by labelling:

       (i)   a lone pair

       (ii) a covalent bond.

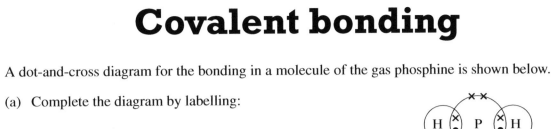

(2 marks)

   (b) Give the formula of phosphine.

   .... **(1 mark)**

   (c) Is phosphine a compound or an element?

   .... **(1 mark)**

2 The table below shows information about different gases in the air. Complete the table.

> The first dot-and-cross diagram has been completed for you. The next two have been started.

Guided

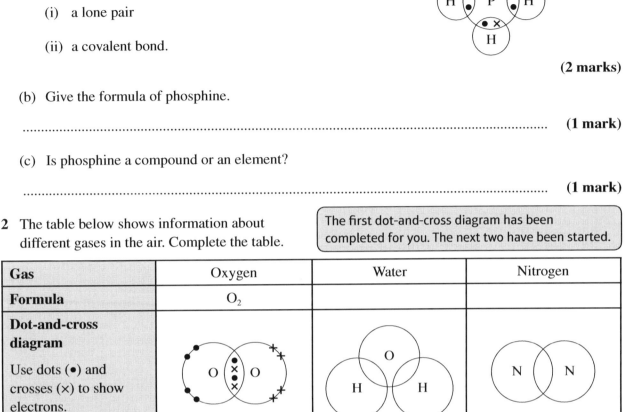

Gas	Oxygen	Water	Nitrogen
Formula	O_2		
Dot-and-cross diagram Use dots (•) and crosses (×) to show electrons. Show only outer-shell electrons.			

(3 marks)

3 Silicon atoms can form covalent bonds with hydrogen atoms to form the compound silane.

   (a) What is a covalent bond?

   .... **(1 mark)**

   (b) Complete the dot-and-cross diagram to show the covalent bonding in a molecule of silane. Show the outer electrons only.

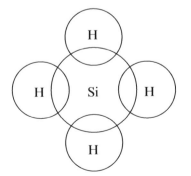

> Silicon is in Group 4, so it has four outer electrons in each atom that it needs to share. Hydrogen has one outer electron in each atom, so four hydrogen atoms each share one electron with each silicon atom.

(2 marks)

Small molecules

1 The molecules of two chlorine compounds are shown in the diagrams below.

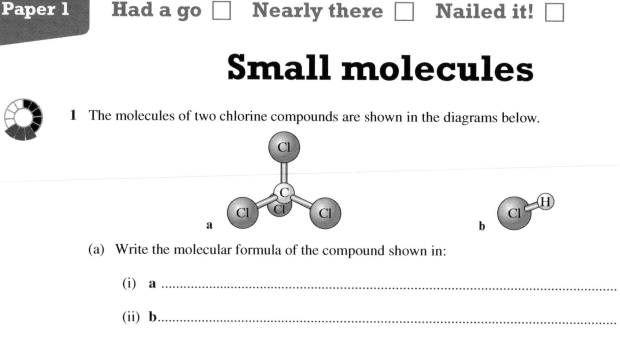

(a) Write the molecular formula of the compound shown in:

(i) **a** .... **(1 mark)**

(ii) **b**.. **(1 mark)**

(b) Draw a diagram of the compound in **a** above using letters to represent the atoms and a line to represent each single bond.

(1 mark)

(c) Draw a dot-and-cross diagram to represent the compound in **b**.

(1 mark)

(d) Why does the compound in **a** have a higher boiling point than the compound in **b**?

Tick **one** box.

☐ It is covalently bonded.

☐ It is ionically bonded.

☐ The covalent bonds between the atoms are stronger.

☐ The forces between the molecules are stronger.   **(1 mark)**

2 Which compound in the table below, A, B, C or D, is a covalent liquid at room temperature and pressure?

	Melting point in °C	Boiling point in °C	Solubility in water	Electrical conductivity of liquid
A	−39	357	poor	good
B	−94	79	poor	poor
C	610	1383	good	good
D	961	2227	poor	good

.... **(1 mark)**

Polymer molecules

1 Circle the correct words to complete the sentences below.

Polymers have very **large/small** molecules.

The **atoms/ions** in the polymer molecules are linked to others by **strong/weak** covalent bonds.

The intermolecular forces between polymer molecules are relatively **strong/weak** so these
substances are solids at room temperature.   **(4 marks)**

2 The structure of poly(ethene) is represented below.

$$\left[\begin{array}{cc} H & H \\ | & | \\ -C & -C- \\ | & | \\ H & H \end{array}\right]_n \nearrow B$$

(a) What does *n* represent?

.... **(1 mark)**

(b) What label should be placed at B?

.... **(1 mark)**

> **Guided**

(c) Show in a diagram three units of poly(ethene).

$$\begin{array}{cc} H & H \\ | & | \\ -C & -C- \\ | & | \\ H & H \end{array}$$

> One repeating unit has
> been drawn for you; draw
> two more.

(1 mark)

(d) Why does poly(ethene) have a much higher melting
point than ethene?

> Remember that it is **not** the covalent
> bonds that break when a molecule melts.

....

.... **(2 marks)**

3 PVC is a polymer used in window frames and guttering. Its structure is shown below.

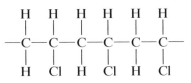

Draw the repeating unit in this polymer.

(1 mark)

Diamond and graphite

1 Many substances have giant covalent structures or simple molecular structures.

Choose the correct structure, giant covalent or simple molecular, for each of the following substances.

(a) ammonia .... **(1 mark)**

(b) carbon dioxide .... **(1 mark)**

(c) silicon dioxide... **(1 mark)**

(d) water .... **(1 mark)**

> You only need to know of three giant covalent structures: graphite, diamond and silicon dioxide.

2 Diamond and silicon dioxide are both very hard solids with high melting points.

(a) Explain why silicon dioxide has a very high melting point.

....

.... **(3 marks)**

(b) Describe the structure of diamond.

> In your answer use these words: carbon atoms, strong, giant covalent, four, covalent bonds.

.... **(3 marks)**

3 When graphite melts, what type of bonds are broken?

Tick **one** box.

☐ covalent bonds   ☐ metallic bonds

☐ ionic bonds   ☐ intermolecular bonds   **(1 mark)**

4 The diagram below shows the structure of graphite. Complete the diagram by inserting the correct labels.

A ....

B ....

C ....

(3 marks)

5 Why does graphite conduct electricity?

Tick **one** box.

☐ It contains carbon atoms that are free to move.

☐ It contains delocalised electrons that are free to move.

☐ It has weak intermolecular forces between the layers.

☐ It contains layers that slide and move over each other.   **(1 mark)**

Graphene and fullerenes

1 Which structure is shown in the diagram?

Tick **one** box.

☐ buckminsterfullerene

☐ diamond

☐ graphene

☐ graphite

each carbon atom joined to three others

strong covalent bond

(1 mark)

2 Fullerenes are molecules of carbon atoms arranged in rings.

(a) How many carbon atoms do most rings of a fullerene contain? Choose any two numbers from the box.

4	5	6	7	8	20	60

.... **(2 marks)**

> **Guided**

(b) Name **two** fullerenes.

carbon nanotubes and .... **(2 marks)**

(c) Suggest why fullerenes are used in drug delivery systems in the body.

....

.... **(2 marks)**

3 Describe the difference in structure between graphite and graphene.

....

.... **(2 marks)**

4 The diagram shows the structure of a cylindrical fullerene.

(a) Name this type of fullerene.

.... **(1 mark)**

(b) Complete the sentence below about the fullerene in the diagram.

This fullerene is a .... conductor of electricity and has a .... tensile strength.

(2 marks)

(c) Suggest why the fullerene in the diagram can be used as a lubricant.

....

.... **(2 marks)**

Metallic bonding

1 The diagram below shows the structure of the metal sodium.

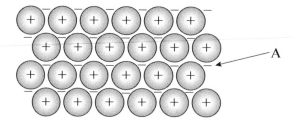

(a) Name particle A. .... **(1 mark)**

(b) Describe the structure of the metal sodium. .... **(2 marks)**

....

(c) What is a metallic bond? .... **(2 marks)**

....

2 The table below gives some properties of the metal calcium and one of its compounds, calcium chloride.

| | Substance | |
Property	Calcium	Calcium chloride
Melting point in °C	842	772
Electrical conductivity when solid	conducts	does not conduct
Electrical conductivity when molten	conducts	conducts

(a) Name the type of bonding found in

> Remember, there are three types of strong bonding: metallic, ionic and covalent.

(i) calcium .... **(1 mark)**

(ii) calcium chloride. .... **(1 mark)**

(b) Use ideas about structure and bonding to explain the similarities and differences between the properties of calcium and calcium chloride.

> **Guided**

Calcium conducts when both solid and molten, but calcium chloride only conducts when molten. The conductivity of solid calcium is due to the presence of delocalised electrons that can move and carry charge. Solid calcium chloride ....

....

....

....

....

....

.... **(6 marks)**

Giant metallic structures and alloys

1　Gold (Au) is a malleable metal. This means it can be easily hammered into shape.

(a)　(i)　Explain why gold is malleable.

....

.... **(2 marks)**

> **Guided**

(ii)　Give the name and the number present of each particle found in the nucleus of a gold atom. You may find the periodic table on page 122 useful.

The number of protons equals the number of electrons in an atom, and equals the atomic number.

Protons ....

.... **(2 marks)**

The mass number minus the atomic number equals the number of neutrons.

(b)　Pure gold is very soft. It is often alloyed with other metals, such as copper or silver, for use in jewellery.

Explain, in terms of its structure, why an alloy is harder than a pure metal.

....

....

.... **(2 marks)**

(c)　Gold is a very unreactive metal. However, it will react slowly with chlorine gas at room temperature to form gold(III) chloride, $AuCl_3$.

Remember that chlorine has the formula Cl_2.

Write a balanced symbol equation for this reaction.

.... **(2 marks)**

2　Why are metals good conductors of thermal energy?

Tick **one** box.

☐　Metals contain delocalised electrons, which can move and transfer energy.

☐　Metals contain ions that can move and transfer energy.

☐　Metals have layers in their structure that allow energy transfer.

☐　Metals have ionic bonding, which allows energy transfer.　　**(1 mark)**

The three states of matter

1 The diagram below shows a simple model of a solid. Small spheres are used to represent particles.

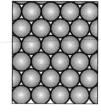

(a) Describe the movement and arrangement of the particles in a solid.

.... **(2 marks)**

(b) Describe what happens to the movement and arrangement of the particles when the solid is heated until it changes to a liquid.

....

.... **(2 marks)**

(c) Give **two** limitations of representing the particles as shown in the diagram above.

....

.... **(2 marks)**

2 The table below shows some physical properties of four substances, A, B, C and D.

Substance	Melting point in °C	Boiling point in °C	Electrical conductivity as solid	Electrical conductivity as liquid
A	−95	69	poor	poor
B	325	1755	good	good
C	800	1412	poor	good
D	113	184	poor	poor

> **Guided**

(a) Which substance is

> It is a good conductor of electricity and has high melting and boiling points.

> Sodium chloride is ionic so it conducts only when molten.

   (i) a metal

   B is a metal. **(1 mark)**

   (ii) sodium chloride

.... **(1 mark)**

   (iii) a covalent solid at room temperature?

> Room temperature is about 20 °C, so solids at room temperature must have a melting point and boiling point above this. A covalent solid does not conduct electricity.

.... **(1 mark)**

(b) Explain why the melting point of C is high and that of A is low.

....

....

.... **(4 marks)**

Nanoscience

1 Nanoscience is a branch of research into the properties of nanoparticles, which have applications in a number of different areas.

(a) What is a nanoparticle?

> In your answer refer to the size of the particle.

.... **(1 mark)**

> **Guided**

(b) Describe how the surface area of nanoparticles compares with that of ordinary powder particles.

Compared with the particles in ordinary powders, nanoparticles have a surface

area that is very .... **(1 mark)**

2 Read the information in the box and then answer the questions.

> **Nanotechnology, good or bad?**
> Nanoparticles with many interesting new properties have been developed, and these are used in sunscreens, drug delivery, catalysts and computing. However, some scientists are concerned about the introduction of new nanotechnology. As they are so small, nanoparticles can get everywhere and can be absorbed into any part of the human body. Their full effects are unpredictable as we don't yet know all their properties. For example, silver nanoparticles, which can be used in place of antibiotics to kill bacteria, might damage other cells in different parts of the body. We therefore have to do more research, and control the trials and use of this new technology very carefully.

(a) Give **one** reason why nanoparticles are potentially so useful.

....

.... **(1 mark)**

(b) What general property of nanoparticles makes them potentially useful as catalysts?

.... **(1 mark)**

(c) Why do we have to control the introduction of nanotechnology?

.... **(1 mark)**

3 (a) Give the diameter of an atom and of a nanoparticle in metres.

> **Maths skills** Remember that the radius of an atom is 1×10^{-10} m, so the diameter is twice this.

....

....

.... **(2 marks)**

> **Guided**

(b) What is the difference in surface-area-to-volume ratio between a particle of diameter 1×10^{-8} m and a particle of diameter 1×10^{-9} m?

The surface-area-to-volume ratio is ....

by a factor of .... **(1 mark)**

Extended response – Bonding and structure

Copper is a transition metal. Some of the physical properties of copper are shown in the table below.

Physical properties of copper
high melting point
good conductor of electricity
good conductor of heat
soft and malleable

Explain the physical properties of copper in terms of structure and bonding.

> Before you begin, think about the type of bonding present in copper and the type of particles that make up the structure. Make sure you use the correct terminology in your answer.

....

....

....

....

....

....

....

....

....

....

....

....

....

....

....

....

....

(6 marks)

> When you have finished your answer, read it through and tick each of the physical properties you have explained to ensure you have not left any out.

Relative formula mass

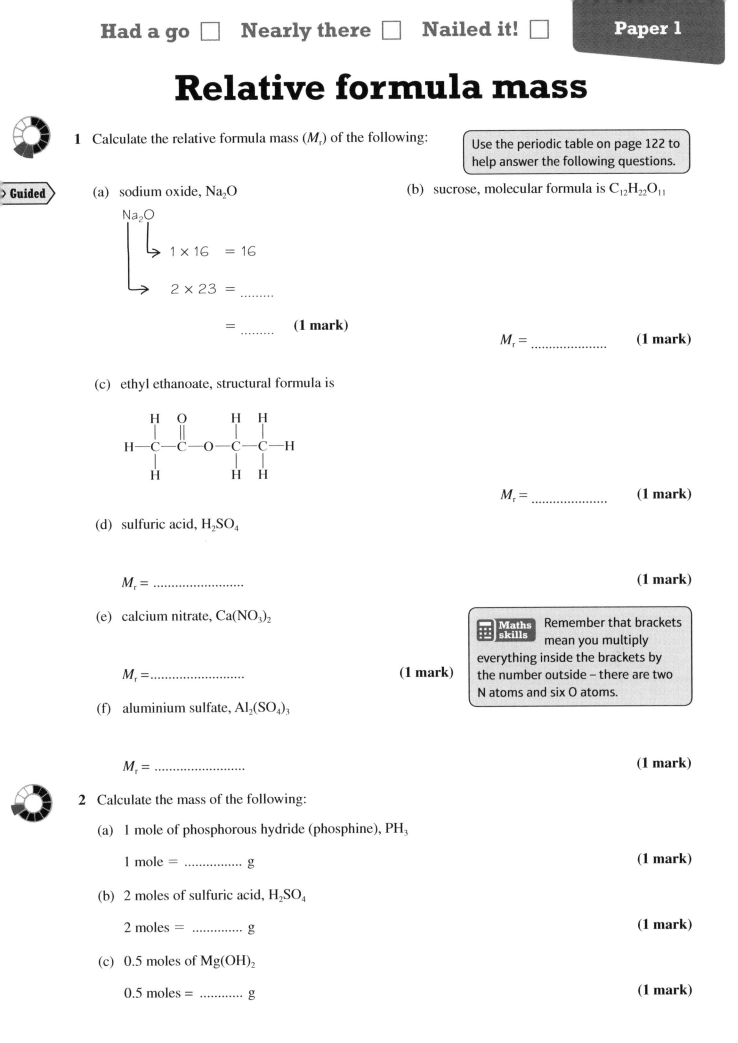

1 Calculate the relative formula mass (M_r) of the following:

Use the periodic table on page 122 to help answer the following questions.

> **Guided**

(a) sodium oxide, Na_2O

Na_2O

$1 \times 16 \ = 16$

$2 \times 23 \ = $ ....

$= $ .... **(1 mark)**

(b) sucrose, molecular formula is $C_{12}H_{22}O_{11}$

$M_r = $ .... **(1 mark)**

(c) ethyl ethanoate, structural formula is

$$\begin{array}{c} H \quad O \qquad H \quad H \\ | \quad\quad || \qquad | \quad\ | \\ H-C-C-O-C-C-H \\ | \qquad\qquad | \quad\ | \\ H \qquad\qquad H \quad H \end{array}$$

$M_r = $ .... **(1 mark)**

(d) sulfuric acid, H_2SO_4

$M_r = $ .... **(1 mark)**

(e) calcium nitrate, $Ca(NO_3)_2$

$M_r = $...................... **(1 mark)**

Maths skills Remember that brackets mean you multiply everything inside the brackets by the number outside – there are two N atoms and six O atoms.

(f) aluminium sulfate, $Al_2(SO_4)_3$

$M_r = $ .... **(1 mark)**

2 Calculate the mass of the following:

(a) 1 mole of phosphorous hydride (phosphine), PH_3

1 mole = ... g **(1 mark)**

(b) 2 moles of sulfuric acid, H_2SO_4

2 moles = ... g **(1 mark)**

(c) 0.5 moles of $Mg(OH)_2$

0.5 moles = ... g **(1 mark)**

Moles

1 What is the number of atoms in 5.6 g of iron?

> One mole of iron has a mass of 56 g. Remember that in one mole there are 6.02×10^{23} atoms of iron. You can be sure that the first answer is wrong because it is much, much smaller than a mole.

Guided

Tick **one** box.

☐ ~~5.60×10~~   ☐ 5.60×10^{23} ☐ 6.02×10^{22} ☐ 6.02×10^{24}

(1 mark)

2 Calculate the number of moles in the following:

(a) 88 g of carbon dioxide, CO_2

Number of moles = ... **(2 marks)**

(b) 4 g of oxygen, O_2

Number of moles = ... **(2 marks)**

(c) 7.4 g of calcium hydroxide, $Ca(OH)_2$

Number of moles = ... **(2 marks)**

(d) 17.1 g of aluminium sulfate, $Al_2(SO_4)_3$

Number of moles = ... **(2 marks)**

3 Calculate the mass of the following:

Guided

(a) 0.25 moles of $CaCO_3$

Relative formula mass of $CaCO_3 = 40 + 12 + (3 \times 16) = 100$

Mass of substance = number of moles × relative formula mass =

0.25 moles = ... g **(2 marks)**

(b) 1.2 moles of $Mg(NO_3)_2$

1.2 moles = ... g **(2 marks)**

Balanced equations, moles and masses

1 A tablet for indigestion contains sodium hydrogen carbonate and citric acid. When added to water the tablet fizzes as the sodium hydrogen carbonate reacts with the citric acid to produce a salt, called a citrate.

(a) (i) Complete the word equation for the formation of the citrate salt from citric acid.

sodium hydrogen carbonate + citric acid →

.... + ... + ... **(1 mark)**

(ii) Use the equation in (a) (i) to explain what causes the fizz.

.... **(1 mark)**

(b) An antacid tablet is added to 50 cm³ of water in a conical flask. The flask is loosely stoppered with a cotton-wool plug and placed on a balance. The initial reading is 103.261 g.

> **Guided**

(i) Draw a labelled diagram of the apparatus.

> The balance has been drawn for you. The description mentions the other things you need to draw.

103.261 g —top-pan balance

(3 marks)

(ii) Explain why the total mass of the flask and contents decreases during the experiment.

....

.... **(2 marks)**

(iii) What is the purpose of the cotton wool plug?

.... **(1 mark)**

(iv) The experiment is repeated using calcium oxide and citric acid. Explain what happens to the balance reading during this reaction.

....

.... **(2 marks)**

2 Magnesium and hydrochloric acid react as shown in this equation:

> **Maths skills** Remember that two moles of HCl are used so $2 \times M_r$ is needed.

$$Mg + 2HCl \rightarrow MgCl_2 + H_2$$

Using relative formula masses (see the periodic table on page 122), show that this equation is balanced.

(2 marks)

3 Some magnesium was weighed in a crucible, covered with a lid and heated. The results are below.

crucible + lid = 30.00 g   magnesium + crucible + lid = 30.24 g

magnesium oxide + crucible + lid = 30.40 g

Calculate the mass of:

(a) magnesium used .... **(1 mark)**

(b) magnesium oxide produced .... **(1 mark)**

(c) oxygen used. .... **(1 mark)**

Reacting masses

1 1.2 g of magnesium is added to a solution of copper(II) sulfate. The equation for the reaction that occurs is:

$$Mg + CuSO_4 \rightarrow Cu + MgSO_4$$

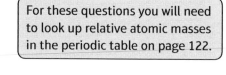

For these questions you will need to look up relative atomic masses in the periodic table on page 122.

What mass of copper is formed?

Tick **one** box.

☐ 1.2 g   ☐ 3.2 g

☐ 2.4 g   ☐ 6.4 g   **(1 mark)**

Guided

2 Copper nitrate crystals are made when copper carbonate reacts with nitric acid. The equation for the reaction is:

$$CuCO_3 + 2HNO_3 \rightarrow Cu(NO_3)_2 + H_2O + CO_2$$

Calculate the mass of copper carbonate needed to react with dilute nitric acid to make 9.4 g of copper nitrate.

Relative formula mass of $Cu(NO_3)_2$ = 63.5 + 2 × (14 + [3 × 16]) = 187.5

Moles of copper nitrate = $\dfrac{9.4}{187.5}$ = 0.05

Ratio $1CuCO_3 : 1Cu(NO_3)_2$

So, 0.05 moles of $CuCO_3$ are required.

You now need to work out the formula mass of $CuCO_3$ and use moles × M_r to find the mass.

Mass of copper carbonate = ... g   **(4 marks)**

3 5.53 g of potassium manganate(VII), $KMnO_4$, is heated to constant mass. The following reaction occurs:

$$2KMnO_4 \rightarrow K_2MnO_4 + MnO_2 + O_2$$

Calculate the mass of oxygen, O_2, that forms when 5.53 g of potassium manganate(VII) is heated.

Mass of oxygen = ... g   **(4 marks)**

4 Iron is produced in the blast furnace by reduction of iron(III) oxide. The equation for the reaction is:

$$Fe_2O_3 + 3CO \rightarrow 3CO_2 + 2Fe$$

Calculate the mass of iron, in grams, which is produced when 10 kg of iron(III) oxide is reduced.

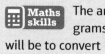

Maths skills The answer is needed in grams, so your first step will be to convert 10 kg into grams.

Mass of iron = ... g   **(4 marks)**

Further reacting masses

1   Chlorine reacts with ethanoic acid as shown by this equation:

$$3Cl_2 + CH_3COOH \rightarrow CCl_3COOH + 3HCl$$

10.65 g of chlorine is reacted with 6.0 g of CH_3COOH.

> For these questions you will need to look up relative atomic masses in the periodic table on page 122.

(a) Calculate

(i)   the number of moles of chlorine in 10.65 g

Number of moles = ...   **(2 marks)**

(ii) the number of moles of CH_3COOH in 6.0 g

Number of moles = ...   **(2 marks)**

(b) Use the ratio of the equation to determine the limiting reactant.

> Remember the limiting reactant is the one that will be completely used up and determines how much product forms.

....**(2 marks)**

(c) Calculate the mass of CCl_3COOH formed.

Mass = ... g   **(2 marks)**

2   Ammonium chloride reacts with calcium oxide according to this equation:

$$2NH_4Cl + CaO \rightarrow CaCl_2 + H_2O + 2NH_3$$

Guided

Calculate the mass of calcium chloride obtained from 20.0 g of ammonium chloride and 50.0 g of calcium oxide. Give your answer to three significant figures.

$$\text{Moles } NH_4Cl = \frac{20}{53.5} = 0.374$$

> First calculate the number of moles of ammonium chloride and calcium oxide.

$$\text{Moles } CaO = \frac{50}{56} = 0.893$$

> Then write down the ratio and use it to find out which reactant is in excess.

$$2 \text{ mol } NH_4Cl : 1 \text{ mol } CaO$$

$$0.374 : \frac{0.374}{2}$$

.... is in excess and .... is the

limiting reactant.

> Use the number of moles of the limiting reactant to find the number of moles of calcium chloride, and then use moles × M_r to find the mass of calcium chloride.

Mass = ... g   **(5 marks)**

Concentration of a solution

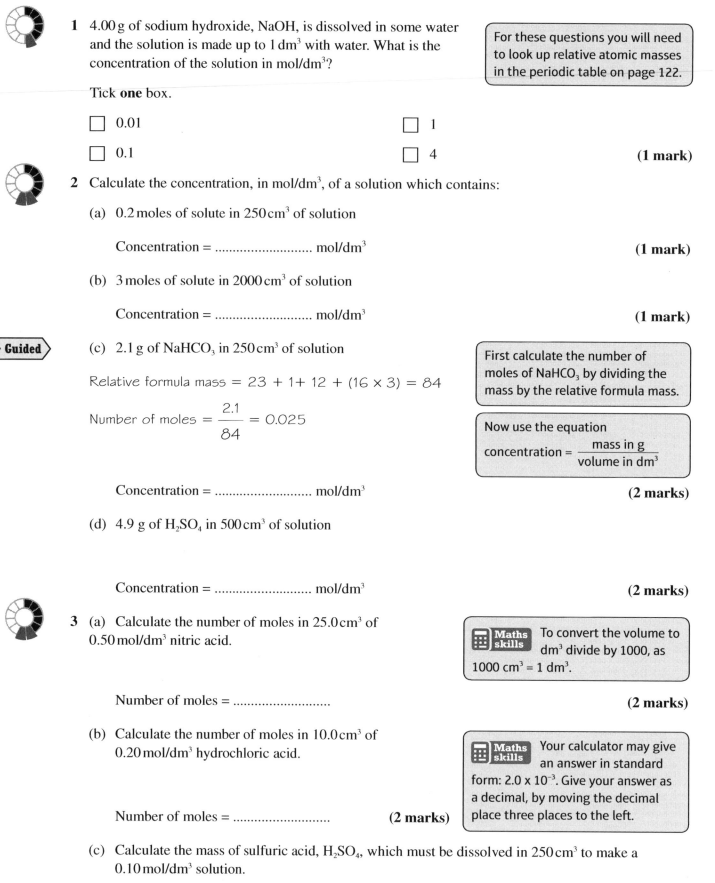

1 4.00 g of sodium hydroxide, NaOH, is dissolved in some water and the solution is made up to 1 dm³ with water. What is the concentration of the solution in mol/dm³?

> For these questions you will need to look up relative atomic masses in the periodic table on page 122.

Tick **one** box.

☐ 0.01   ☐ 1

☐ 0.1   ☐ 4   **(1 mark)**

2 Calculate the concentration, in mol/dm³, of a solution which contains:

(a) 0.2 moles of solute in 250 cm³ of solution

Concentration = ... mol/dm³   **(1 mark)**

(b) 3 moles of solute in 2000 cm³ of solution

Concentration = ... mol/dm³   **(1 mark)**

⟩ **Guided** ⟩   (c) 2.1 g of NaHCO₃ in 250 cm³ of solution

Relative formula mass = 23 + 1 + 12 + (16 × 3) = 84

Number of moles = $\dfrac{2.1}{84}$ = 0.025

> First calculate the number of moles of NaHCO₃ by dividing the mass by the relative formula mass.

> Now use the equation
> concentration = $\dfrac{\text{mass in g}}{\text{volume in dm}^3}$

Concentration = ... mol/dm³   **(2 marks)**

(d) 4.9 g of H₂SO₄ in 500 cm³ of solution

Concentration = ... mol/dm³   **(2 marks)**

3 (a) Calculate the number of moles in 25.0 cm³ of 0.50 mol/dm³ nitric acid.

> **Maths skills** To convert the volume to dm³ divide by 1000, as 1000 cm³ = 1 dm³.

Number of moles = ...   **(2 marks)**

(b) Calculate the number of moles in 10.0 cm³ of 0.20 mol/dm³ hydrochloric acid.

> **Maths skills** Your calculator may give an answer in standard form: 2.0 x 10⁻³. Give your answer as a decimal, by moving the decimal place three places to the left.

Number of moles = ...   **(2 marks)**

(c) Calculate the mass of sulfuric acid, H₂SO₄, which must be dissolved in 250 cm³ to make a 0.10 mol/dm³ solution.

Mass = ... g   **(2 marks)**

Practical skills

Core practical – Titration

1 A student carried out a titration to find the volume of 0.100 mol/dm³ hydrochloric acid needed to neutralise 25.0 cm³ of a potassium hydroxide solution of unknown concentration. The apparatus was set up as shown in the diagram on the right.

A ....

0.100 mol/cm³ HCl solution

B ....

25.0 cm³ of KOH solution

C ....

(a) Complete the diagram by writing the correct labels at positions A, B and C. **(3 marks)**

(b) What piece of apparatus should be used to place 25.0 cm³ of potassium hydroxide solution into the piece of apparatus labelled B?

.... **(1 mark)**

(c) Describe how the titration is carried out. Include the name of a suitable indicator, and give the colour change that would be seen.

....

....

....

.... **(4 marks)**

(d) What is the purpose of the piece of apparatus labelled C?

.... **(1 mark)**

> **Guided**

(e) Some results were recorded in the table below.

	Titration 1	Titration 2	Titration 3	Titration 4
End volume in cm³	27.25	26.85	26.10	26.40
Start volume in cm³	0.05	0.10	0.05	0.25
Titre in cm³	27.25 − 0.05 = 27.20			

(i) Complete the table by calculating the titre values. **(4 marks)**

> Use concordant values which are within 0.10 cm³ of each other.

(ii) Calculate the mean titre.

> **Maths skills** To calculate the mean, add the results and divide by the total number of results.

.... **(2 marks)**

(iii) Using the mean titre and the equation for the reaction, calculate the concentration of the potassium hydroxide solution.

$$\text{Moles} = 26.10 \times \frac{0.100}{1000} = \text{.............................}$$

HCl(aq) + KOH(aq) → KCl(aq) + H₂O(l); ratio is 1 mol HCl : 1 mol KOH

So amount of KOH reacting = ...

amount of KOH in 25 cm³ of solution in the flask, is ....

Concentration = ... mol/dm³ **(5 marks)**

Titration calculations

Guided

1 Dilute nitric acid and lithium hydroxide solution react as shown by this equation:

$$HNO_3 + LiOH \rightarrow LiNO_3 + H_2O$$

5.0 cm³ of nitric acid was used to neutralise exactly 15.0 cm³ of 0.20 mol/dm³ lithium hydroxide solution.

Calculate the concentration of the nitric acid.

Moles of lithium hydroxide = volume in cm³ × $\dfrac{\text{concentration in mol/dm}^3}{1000}$

$= 15.0 \times \dfrac{0.2}{1000} = 0.003$ mol

1 mole of LiOH reacts with 1 mole of HNO_3

0.003 moles of LiOH react with .... moles of HNO_3

> Use the equation moles of HNO_3 × 1000/volume in cm³ to find the concentration.

Concentration = ... mol/dm³   **(5 marks)**

2 25.0 cm³ of sodium hydroxide solution is exactly neutralised by 7.5 cm³ of 1.0 mol/dm³ hydrochloric acid. The equation for the reaction is:

$$HCl(aq) + NaOH(aq) \rightarrow NaCl(aq) + H_2O(l)$$

Calculate the concentration of the sodium hydroxide in mol/dm³.

Concentration = ... mol/dm³   **(3 marks)**

3 25.0 cm³ of sodium carbonate solution reacts with 25.6 cm³ of 0.100 mol/dm³ hydrochloric acid solution. The equation for the reaction is:

$$Na_2CO_3 + 2HCl \rightarrow 2NaCl + H_2O + CO_2$$

Calculate the concentration of the sodium carbonate solution.
Give your answer to two significant figures.

> 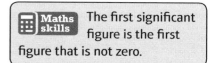 **Maths skills** The first significant figure is the first figure that is not zero.

Concentration = ... mol/dm³   **(4 marks)**

4 Sodium hydroxide solution reacts with dilute sulfuric acid as shown by this equation:

$$2NaOH + H_2SO_4 \rightarrow Na_2SO_4 + H_2O$$

Which one of the following solutions of sodium hydroxide will exactly neutralise 20.0 cm³ of 1.0 mol/dm³ sulfuric acid?

> Remember to work out the number of moles of acid, and use the ratio.

Tick **one** box.

☐ 20.0 cm³ of concentration 0.5 mol/dm³   ☐ 20.0 cm³ of concentration 1.0 mol/dm³

☐ 40.0 cm³ of concentration 0.5 mol/dm³   ☐ 20.0 cm³ of concentration 2.0 mol/dm³

(1 mark)

Reactions with gases

1 Calculate the volume of 34 g of chlorine gas, Cl_2, at room temperature and pressure.

> The volume of one mole of any gas at room temperature (20 °C and 1 atmosphere pressure) is 24 dm³.

Volume of chlorine = ... dm³   **(2 marks)**

2 Calcium reacts with 25 cm³ of 0.10 mol/dm³ of hydrochloric acid as shown by this equation:

$Ca + 2HCl \rightarrow CaCl_2 + H_2$

Calculate the volume of hydrogen gas produced at room temperature and pressure.

Volume of hydrogen = ... dm³   **(4 marks)**

3 100 cm³ of propane gas burns completely in 500 cm³ of oxygen. The equation is:

$C_3H_8(g) + 5O_2(g) \rightarrow 3CO_2(g) + 4H_2O(g)$

Calculate the volume of:

> The reaction only involves gases, so the volumes are in the same ratio as the moles in the equation.

(a) carbon dioxide produced

Volume of carbon dioxide = ... cm³   **(1 mark)**

(b) steam produced

Volume of steam = ... cm³   **(1 mark)**

4 Calcium carbonate reacts with hydrochloric acid as shown in this equation:

$CaCO_3 + 2HCl \rightarrow CO_2 + CaCl_2 + H_2O$

20 cm³ of 0.2 mol/dm³ hydrochloric acid is reacted with calcium carbonate.

> **Guided**

(a) Calculate the volume of carbon dioxide produced. Give your answer in cm³.

Moles of hydrochloric acid = $20 \times \dfrac{0.2}{1000}$ =

> Now use the ratio to find the moles of carbon dioxide, and then use this equation:
>
> Moles of carbon dioxide = $\dfrac{\text{moles of gas}}{24\,000}$

> **Maths skills** To convert from dm³ to cm³, multiply by 1000.

Volume of carbon dioxide = ... cm³   **(5 marks)**

(b) Calculate the mass of calcium carbonate needed to react with all the hydrochloric acid.

Mass of calcium carbonate = ... g   **(3 marks)**

Reaction yields

1 A student was investigating the formation of magnesium oxide by burning magnesium in air.

The balanced equation for the reaction is:

$2Mg + O_2 \rightarrow 2MgO$

The student calculated that 1.2 g of magnesium should react to produce 2.0 g of magnesium oxide.

(a) What mass of oxygen would combine with 1.2 g of magnesium to produce 2.0 g of magnesium oxide?

.... **(1 mark)**

> Use the law of conservation of mass to check your answer.

(b) If the experiment had a 50% yield, how much magnesium oxide would be obtained?

.... **(1 mark)**

(c) Suggest the **two** most likely reasons for the low yield of magnesium oxide. In each case explain your reason with reference to the experimental set-up shown in the diagram.

....

.... **(2 marks)**

2 In the preparation of copper nitrate, some copper carbonate was reacted with nitric acid. The equation for the reaction is:

$CuCO_3 + 2HNO_3 \rightarrow Cu(NO_3)_2 + H_2O + CO_2$

10.0 g of copper carbonate was reacted with nitic acid.

> The theoretical yield is the maximum mass of product which can be produced. To find this you need to carry out a reacting mass calculation, similar to those on page 32.

> **Guided**

(a) Calculate the theoretical yield, in grams, of copper nitrate.

$$\text{Moles of copper carbonate} = \frac{\text{mass}}{\text{relative formula mass}} = \frac{10.0}{123.5} = 0.08$$

1 mole of copper carbonate produces 1 mole of copper nitrate.

Moles of copper nitrate = ...

Moles of copper nitrate × relative formula mass = mass of copper nitrate

.... × .... = ... **(5 marks)**

(b) 8.0 g of crystals were obtained. Use your answer to part (a) to calculate the percentage yield.

.... **(2 marks)**

3 5.0 g of butan-1-ol (relative formula mass 74) reacted with an excess of hydrogen bromide, and 6.4 g of bromobutane (relative formula mass 137) was obtained after purification.

$C_4H_9OH + HBr \rightarrow C_4H_9Br + H_2O$

Calculate the percentage yield of bromobutane.

Tick **one** box.

☐ 42%   ☐ 54%   ☐ 69%   ☐ 78%

(1 mark)

Atom economy

1 (a) Write an equation for percentage atom economy of a desired product in a reaction.

....

.... **(1 mark)**

(b) Calculate the atom economy for making hydrogen by reacting coal with steam. The carbon dioxide leaves the process and enters the atmosphere as a waste gas:

$$C(s) + 2H_2O(g) \rightarrow CO_2(g) + 2H_2(g)$$

Atom economy = ... %   **(2 marks)**

(c) How could the atom economy for this process be improved?

.... **(1 mark)**

2 Alkanes can be cracked to form alkenes. Decane can be cracked to form two products as shown in these equations:

decane → ethene + octane

$$C_{10}H_{22} \rightarrow C_2H_4 + C_8H_{18}$$

(a) Calculate the atom economy for the production of ethene as the only useful product.

Atom economy = ... %   **(2 marks)**

(b) If both products can be sold, what is the atom economy?

.... **(1 mark)**

> **Guided**

(c) Explain why a reaction with a high atom economy is more economical and more sustainable.

A reaction with a high atom economy is more economical because less material is wasted and so more of the raw materials bought can be sold on. There are fewer disposal costs

for ....

.... **(2 marks)**

3 Copper can be extracted by heating copper(II) oxide with carbon according to this equation:

$$2CuO + C \rightarrow 2Cu + CO_2$$

What is the atom economy for the extraction of copper, assuming that the carbon dioxide is a waste product?

> Remember that two moles of copper are produced, so this has to be included in the calculation.

Tick **one** box.

☐ 40.0%   ☐ 69.5%   ☐ 74.3%   ☐ 80.0%

(1 mark)

Exam skills –
Quantitative chemistry

1 Calcium is a Group 2 metal that reacts with nitric acid. The equation for the reaction is:

$$Ca + 2HNO_3 \rightarrow Ca(NO_3)_2 + H_2$$

In an experiment, hydrogen gas was collected at room temperature and pressure when 0.23 g of calcium metal was reacted with excess nitric acid.

> The volume of one mole of any gas at room temperature and 1 atmosphere pressure is 24.0 dm³.

(a) Calculate the volume of hydrogen gas produced. Give your answer to two significant figures.

> You will need to use moles
> $$\text{of gas} = \frac{\text{volume (dm}^3)}{24}$$

Volume of hydrogen gas = ... dm³   **(6 marks)**

(b) Calculate the volume of 0.2 mol/dm³ nitric acid needed to completely react with the 0.23 g of calcium used.

Volume of nitric acid = ... cm³   **(3 marks)**

Reactivity series

1   The table below shows the results recorded when a piece of metal was placed in a metal ion solution.

Metal	zinc	zinc	copper	copper
Solution	copper sulfate	lead nitrate	lead nitrate	silver nitrate
Result	reaction	reaction	no reaction	reaction

Which list gives the correct order of reactivity for the four metals, starting with the most reactive?

Tick **one** box.

☐ zinc, copper, lead, silver   ☐ silver, copper, zinc, lead

☐ zinc, lead, copper, silver   ☐ lead, zinc, copper, silver   **(1 mark)**

2   Complete the word equations.

›Guided›

(a) magnesium + water → magnesium hydroxide + ... **(1 mark)**

(b) calcium + nitric acid → calcium nitrate + ... **(1 mark)**

(c) zinc + hydrochloric acid → .... + ... **(1 mark)**

3   Potassium reacts with water at room temperature to produce an alkaline solution and a gas.

(a) Name the gas produced. .... **(1 mark)**

(b) Which ion causes the solution to be alkaline? .... **(1 mark)**

(c) What happens to the potassium atoms in this reaction, in terms of their electrons?

.... **(1 mark)**

(d) Lithium also reacts with water at room temperature. State why this reaction is less vigorous than that of potassium and water.

| Answer this by comparing the tendency of the two metals to form positive ions. |

....

.... **(2 marks)**

4   In an experiment some metals were placed into metal salt solutions and any reaction that occurred was recorded in the table below.

| Look at the results for each metal and note how many solutions it reacts with: the more ticks there are, the more reactive the metal. |

Solution / Metal	Copper sulfate	Zinc sulfate	Iron sulfate	Lead nitrate	Tin chloride
Copper		✗	✗	✗	✗
Zinc	✓		✓	✓	✓
Iron	✓	✗		✓	✓
Lead	✓	✗	✗		✗
Tin	✓	✗	✗	✓	

✓ means a reaction occurred   ✗ means a reaction did not occur

Use the results in the table to put the metals in order from the most reactive to the least reactive.

.... **(2 marks)**

Oxidation

1 Each of the following reactions involves oxidation:

Reaction 1: $ZnO + H_2 \rightarrow Zn + H_2O$

Reaction 2: $2Ca + O_2 \rightarrow 2CaO$

Reaction 3: $Mg + FeSO_4 \rightarrow Fe + MgSO_4$

(a) Write the formula of the substance that is oxidised in reaction 1.

.... **(1 mark)**

Guided (b) Explain why reaction 2 involves oxidation.

Calcium has .... oxygen. .... of oxygen
is oxidation.

(2 marks)

(c) Explain, in terms of electrons, why oxidation occurs in reaction 2.

....

.... **(2 marks)**

(d) Write a half-equation for the oxidation in reaction 2.

.... **(1 mark)**

(e) Write an ionic equation for reaction 3.

.... **(2 marks)**

(f) Identify the species that is oxidised in reaction 3.

| 'Species' refers to any element, ion or compound. |

.... **(1 mark)**

2 Fireworks contain, among other things, an oxidiser, a fuel and a colouring agent.

(a) Magnesium is often used as a colouring agent.
Write a symbol equation for burning magnesium.

| Remember, when substances burn they react with oxygen. Oxygen is diatomic. |

.... **(1 mark)**

(b) Carbon is often the fuel in a firework. It burns to form carbon dioxide. Write a symbol
equation for carbon burning and explain why this is an oxidation reaction.

....

.... **(3 marks)**

Guided (c) Oxidisers provide the oxygen needed to allow the
firework to burn effectively. A common oxidiser is
potassium nitrate, which thermally decomposes to
produce potassium oxide, nitrogen and oxygen.
Write a balanced symbol equation for this reaction.

| The formula of potassium nitrate is KNO_3. Work out and add the formulae of potassium oxide, nitrogen and oxygen. Then balance the equation. |

.... $KNO_3 \rightarrow$ .... **(2 marks)**

Reduction and metal extraction

1 Choose an element from the box to answer each of the following questions. Each element may be used once, more than once or not at all.

hydrogen	silver	carbon
magnesium	tin	calcium

(a) Name a metal that is found in the Earth's crust as the uncombined element.

.... **(1 mark)**

(b) Name **two** metals that are likely to be found as compounds.

.... **(2 marks)**

(c) Name an element that is used to reduce metal ores.

.... **(1 mark)**

(d) Name an element that is extracted using electrolysis.

.... **(1 mark)**

2 Most metals are extracted from metal oxides found in rocks called ores. Some metals are found as the uncombined elements.

(a) Why are some metals found as the uncombined elements?

.... **(1 mark)**

(b) In industry, iron is manufactured in the blast furnace as shown by this equation:

$$Fe_2O_3 + 3CO \rightarrow 2Fe + 3CO_2$$

> **Guided**

(i) Explain, in terms of change in oxygen content, why this reaction involves both oxidation and reduction.

In the reaction, CO has .... oxygen to form CO_2.

.... of oxygen is oxidation. Fe_2O_3 has...................................

.... **(4 marks)**

(ii) Write a half-equation for the reduction occurring in this reaction.

.... **(1 mark)**

3 Magnesium reacts with copper(II) sulfate solution.
The balanced symbol equation is: $Mg + CuSO_4 \rightarrow Cu + MgSO_4$

> Remember to check the charge on each ion.

(a) Write an ionic equation for this reaction. .... **(2 marks)**

(b) In this reaction, which ion does not undergo any change? .... **(1 mark)**

(c) Name the species that is oxidised in this reaction. .... **(1 mark)**

(d) Write a half-equation for the oxidation reaction. .... **(1 mark)**

Reactions of acids

1 Indigestion is caused by too much hydrochloric acid in the stomach. Some indigestion remedies contain the insoluble compounds magnesium hydroxide and aluminium hydroxide to react with the excess hydrochloric acid.

(a) What one-word term can be used to describe magnesium hydroxide and aluminium hydroxide?

.... **(1 mark)**

(b) Name the **two** salts formed when magnesium hydroxide and aluminium hydroxide react with the excess hydrochloric acid.

.... **(2 marks)**

2 Complete the table below. **(4 marks)**

> Guided

Acid	Base	Salt
hydrochloric acid	lithium hydroxide	lithium chloride
	calcium oxide	calcium nitrate
sulfuric acid	sodium hydroxide	

First relate the second part of the name of the salt, either chloride, nitrate or sulfate, to the name of the acid used. Then place the name of the metal in front. The first has been completed for you.

3 Complete the word equations.

Remember that oxides and hydroxides are both bases, and both react with acids to give a salt and water.

(a) hydrochloric acid + ... → magnesium chloride + hydrogen **(1 mark)**

(b) sulfuric acid + potassium hydroxide → .... + ... **(1 mark)**

(c) nitric acid + sodium carbonate → .... + ... + ... **(1 mark)**

(d) copper oxide + sulfuric acid → .... + ... **(1 mark)**

4 A student wished to find the volume of hydrochloric acid, HCl, needed to exactly neutralise $20\,cm^3$ of a solution of calcium hydroxide, $Ca(OH)_2$.

(a) Explain what the student should do to find the point when the calcium hydroxide was exactly neutralised.

....

.... **(2 marks)**

(b) Complete the balanced chemical equation for this reaction (state symbols are not required).

.... + ... → .... + $2H_2O$ **(2 marks)**

> Guided

(c) Explain, in terms of electrons, whether calcium is oxidised or reduced when it reacts with hydrochloric acid to form calcium chloride. Write a half-equation for the process.

$Ca →$ ....

The calcium has .... electrons, and .... **(3 marks)**

Practical skills

Core practical – Salt preparation

1 Describe how a sample of cobalt chloride crystals could be made from cobalt oxide and dilute hydrochloric acid.

Guided

Add excess cobalt oxide to a measured quantity of dilute hydrochloric acid in a

beaker and stir. ....

....

....

....

....

.... **(4 marks)**

2 Two groups of students were asked to make some crystals of copper sulfate.

Group 1 added excess copper metal to some dilute sulfuric acid in a beaker. Group 2 added excess solid copper carbonate to some dilute sulfuric acid in a beaker.

(a) Only one group was successful in making the salt. State which of the groups would be unsuccessful in this task and explain why.

....

.... **(2 marks)**

(b) Complete the labels on the diagram to show how the excess copper carbonate would be removed.

(2 marks)

(c) Describe how crystals of copper sulfate can be obtained from the salt solution.

....

....

....

....

.... **(2 marks)**

....

....

....

....

(d) Sodium also produces a salt when it reacts with dilute sulfuric acid.

(i) Name the salt formed between sodium and sulfuric acid. .... **(1 mark)**

(ii) Explain why the addition of sodium metal to sulfuric acid would not be used as a method of preparing sodium sulfate in the laboratory.

> Think about the position of sodium in the reactivity series.

....

.... **(2 marks)**

The pH scale

1   The pH scale is used to measure acid and alkaline properties. The table shows the pH of five solutions.

Solution	A	B	C	D	E
pH	2	6	7	10	13

(a) Which of these solutions contain excess H^+ ions? .... **(1 mark)**

(b) Which solution contains the greatest concentration of OH^- ions? .... **(1 mark)**

(c) How would the pH change if pure water were added to solution C? .... **(1 mark)**

> **Guided** (d) Describe how a student could test the pH of an unknown solution.

The student could add some universal indicator and ....

.... **(2 marks)**

2   Sulfuric acid is a strong acid which neutralises potassium hydroxide.

(a) Write a balanced chemical equation for this reaction.

.... **(2 marks)**

(b) What is meant by a strong acid? | Give your answer in terms of ionisation. |

....

.... **(2 marks)**

(c) Write the ionic equation for a neutralisation reaction. Include state symbols.

.... **(2 marks)**

(d) Solution X has a pH of 4.1. Suggest the pH of solution Y, which has a hydrogen ion concentration ten times lower than that of solution X.

.... **(1 mark)**

3   A is a solution of $2.0\,mol/dm^3$ ethanoic acid and B is a solution of $0.5\,mol/dm^3$ nitric acid.

(a) Which acid, A or B, is a weak acid? .... **(1 mark)**

(b) Which acid, A or B, is more concentrated? Explain your answer.

.... **(2 marks)**

> **Guided** (c) Which acid, A or B, has a lower pH? Explain your answer.

Nitric acid is fully ionised into hydrogen ions in aqueous solution, as it is a

....

.... **(2 marks)**

Electrolysis

1   Molten lead bromide breaks down when it conducts electricity.

    (a)   Using the apparatus shown in the diagram, how would you know if an electric current was passing?

    .... **(1 mark)**

> **Guided**

    (b)   Explain why the lead bromide needs to be molten.

    Solid lead bromide ....

    ....

    ....**(2 marks)**

             graphite electrodes

             lead bromide

    (c)   Name the products at each of the electrodes during this process.

    ....

    .... **(2 marks)**

    (d)   What is a reduction reaction and where does it occur during this process?

       > You need to mention which electrode the different ions move to, and what happens when they get there.

    ....

    .... **(2 marks)**

    (e)   Explain why graphite is used for the electrodes.

    .... **(1 mark)**

2   (a)   Explain what happens to metal ions during electrolysis.

    ....

    ....

    .... **(2 marks)**

    (b)   What is the name of the process that involves loss of electrons?

    .... **(1 mark)**

3   When molten sodium chloride is electrolysed, different reactions occur at the cathode and the anode. Complete the table below.

> **Guided**

> One of the half-equations has been started. Balance it by adding electrons.

	Anode	**Cathode**
Product		
Half-equation	$2Cl^- \rightarrow Cl_2 + \text{.......}$	
Oxidation or reduction?		

                                                   (8 marks)

Aluminium extraction

1 Aluminium is extracted from its ore using electricity. The ore, which mainly contains aluminium oxide, is mixed with cryolite before it is melted and electrolysed.

positive graphite electrode

solid crust

molten mixture of cryolite and aluminium oxide

negative graphite electrode

molten aluminium

(a) Why is cryolite added to the aluminium oxide?

....

....

.... **(2 marks)**

Guided

(b) Why is the aluminium metal formed at the negative electrode?

Aluminium ions are positive ....

....

.... **(2 marks)**

(c) Name **two** products that could be formed at the positive electrode during this process.

.... **(2 marks)**

(d) Complete the half-equation for the reaction that occurs at the negative electrode.

Al^{3+} .... **(2 marks)**

(e) What is the name for the negative electrode?

.... **(1 mark)**

(f) Write the half-equation for the reaction that occurs at the positive electrode.

| Remember that oxygen is diatomic. |

.... **(2 marks)**

(g) Why is electrolysis used to extract aluminium from its ore, instead of reduction with carbon?

....

.... **(1 mark)**

(h) Explain why the positive electrode must be continually replaced.

....

....

....

.... **(3 marks)**

Electrolysis of solutions

1 Why does calcium chloride solution conduct electricity?

   Tick **one** box.

   ☐ It contains electrons that can move.

   ☐ It contains ions that can move.

   ☐ It contains a metal.

   ☐ It contains water.   **(1 mark)**

2 Name the products of electrolysis of potassium bromide solution.

   Tick **one** box.

	Product at the cathode	**Product at the anode**
☐	hydrogen	bromine
☐	hydrogen	oxygen
☐	potassium	bromine
☐	potassium	oxygen

                                                                                 (1 mark)

3 (a) When calcium nitrate solution is electrolysed, the product at the cathode gives a pop
       with a burning splint, and the product at the anode relights a glowing splint.

       (i) Identify the product at the cathode.

       .... **(1 mark)**

       (ii) Identify the product at the anode.

       .... **(1 mark)**

       (iii) Write a half-equation for the reaction at the cathode.

       .... **(2 marks)**

   (b) (i) Complete the table below to show the products at the electrodes during the electrolysis
            of some electrolyte solutions.

Electrolyte solution	Anode	Cathode
copper(II) chloride		
potassium iodide		
sodium bromide		
sodium sulfate		

   > Remember to use the reactivity series. At the cathode, hydrogen is produced if the metal is higher than hydrogen in the series. At the anode, oxygen is produced unless the solution contains halide ions, when the halogen is produced.

                                                                                 (4 marks)

   > Guided

       (ii) What is meant by the word electrolyte?

       An electrolyte is a molten ....

       .... **(1 mark)**

49

Practical skills

Core practical – Electrolysis

1 The diagram opposite shows the apparatus used to electrolyse some aqueous solutions.

(a) Name electrodes A and B and suggest a suitable material for the electrodes.

....

.... **(3 marks)**

(b) Suggest why the aqueous solutions are made up using deionised water rather than tap water.

....

.... **(1 mark)**

electrolyte

electrolyte

A ———— B

power supply

− +

> If a halide ion is present, then a halogen is always produced at the anode.

Guided

(c) The table below gives the results recorded when some aqueous solutions were electrolysed and the products tested. Complete the table to identify the products and write equations for the reactions at the electrodes.

Solution	Potassium chloride	Calcium nitrate	Sulfuric acid	Zinc bromide	Silver nitrate
Observations at cathode	colourless gas	colourless gas	colourless gas	grey solid	white solid
Observations at anode	greenish gas	colourless gas	colourless gas	orange solution	colourless gas
Test used for product at cathode	insert a burning lighted splint result – pop	insert a burning lighted splint result – pop	insert a burning lighted splint result – pop		
Test used for product at anode	universal indicator paper turns red and bleaches		relights a glowing splint	universal indicator paper turns red and bleaches	relights a glowing splint
Identity of product at cathode				zinc	silver
Identity of product at anode				bromine	
Half-equation for reaction at cathode					
Half-equation for reaction at anode					

(23 marks)

(d) Explain **one** important safety instruction that must be followed in this practical.

....

.... **(2 marks)**

Extended response – Chemical changes

Hydrochloric acid reacts with calcium hydroxide solution and with solid calcium. Compare and contrast the reaction of hydrochloric acid with calcium hydroxide solution and the reaction of hydrochloric acid and calcium, in terms of observations, products and equations.

> Think first of the general equations for acids:
>
> - acid + base → salt + water
> - acid + metal → salt + hydrogen
>
> Then write word and balanced equations for the specific reactions.

....

....

....

....

....

....

....

....

....

....

....

....

....

....

....

....

....

....

....

.... **(6 marks)**

Exothermic reactions

Guided

1 What is the law of conservation of energy?

The amount of energy in the Universe at the end of a chemical reaction is

....

.... **(2 marks)**

2 (a) What is an exothermic reaction?

....

....

.... **(2 marks)**

(b) In an exothermic reaction, do the product molecules have more or less energy than the reactants?

.... **(1 mark)**

(c) Name **two** everyday uses of exothermic reactions.

....

.... **(2 marks)**

3 Combustion reactions occur when substances react with oxygen and catch fire.

> Remember that combustion requires oxygen, which is diatomic.

(a) Write a balanced symbol equation for the combustion of methane, CH_4, to produce carbon dioxide and water.

.... **(2 marks)**

(b) What do you expect to happen to the temperature of the surroundings during this reaction?

.... **(1 mark)**

(c) On the axes below, draw a reaction profile for the combustion of methane. Label the reactants, products and axes.

(4 marks)

(d) Draw an arrow on your reaction profile to show the overall energy change during the combustion of methane and label it A.   **(1 mark)**

Endothermic reactions

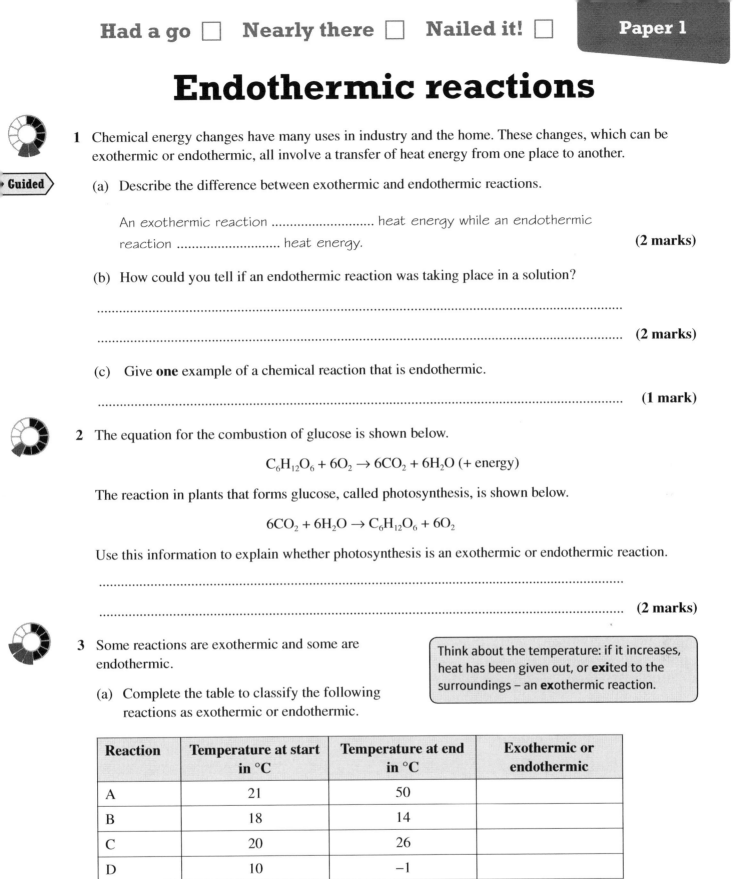

1 Chemical energy changes have many uses in industry and the home. These changes, which can be exothermic or endothermic, all involve a transfer of heat energy from one place to another.

(a) Describe the difference between exothermic and endothermic reactions.

An exothermic reaction .... heat energy while an endothermic

reaction .... heat energy.   **(2 marks)**

(b) How could you tell if an endothermic reaction was taking place in a solution?

....

.... **(2 marks)**

(c) Give **one** example of a chemical reaction that is endothermic.

.... **(1 mark)**

2 The equation for the combustion of glucose is shown below.

$$C_6H_{12}O_6 + 6O_2 \rightarrow 6CO_2 + 6H_2O \text{ (+ energy)}$$

The reaction in plants that forms glucose, called photosynthesis, is shown below.

$$6CO_2 + 6H_2O \rightarrow C_6H_{12}O_6 + 6O_2$$

Use this information to explain whether photosynthesis is an exothermic or endothermic reaction.

....

.... **(2 marks)**

3 Some reactions are exothermic and some are endothermic.

> Think about the temperature: if it increases, heat has been given out, or **exi**ted to the surroundings – an **exo**thermic reaction.

(a) Complete the table to classify the following reactions as exothermic or endothermic.

Reaction	Temperature at start in °C	Temperature at end in °C	Exothermic or endothermic
A	21	50	
B	18	14	
C	20	26	
D	10	−1	

(4 marks)

(b) Calculate the temperature change for Reaction A and for Reaction D.

Temperature change for A = ... °C

Temperature change for D = ... °C   **(2 marks)**

Practical skills

Core practical – Energy changes

1 In an experiment, 25 cm³ of 1 mol/dm³ sodium hydroxide solution was placed, at room temperature (20 °C), in a polystyrene cup, and an excess (40 cm³) of 1 mol/dm³ hydrochloric acid was added. The mixture was stirred with a thermometer and the maximum temperature reached recorded.

thermometer

polystyrene beaker

25 cm³ of sodium hydroxide solution + acid

(a) Why was a polystyrene cup used rather than a glass beaker?

.... **(1 mark)**

(b) Why was the solution stirred after adding the hydrochloric acid?

.... **(1 mark)**

(c) Describe and explain **one** improvement that could be made to the apparatus.

....

.... **(2 marks)**

(d) Calculate the number of moles of sodium hydroxide in the polystyrene cup at the start.

Number of moles = ... **(1 mark)**

(e) The experiment was repeated, adding 40 cm³ of different 1 mol/dm³ acids to 25 cm³ of sodium hydroxide, and the results recorded below.

	Hydrochloric acid	Ethanoic acid	Nitric acid	Sulfuric acid
Maximum temperature in °C	26.0	23.9	25.9	26.1

(i) Explain whether the reaction of sodium hydroxide with an acid is exothermic or endothermic.

....

.... **(2 marks)**

Guided

(ii) Explain how this experiment was a fair test.

The experiment was carried out using the same .... and ....

of acid and the same .... and .... of alkali and was started

at the same .... **(3 marks)**

(iii) Why was the maximum temperature reached lower for ethanoic acid than for the other acids?

....

.... **(3 marks)**

(iv) How could the experiment be changed to show that the maximum temperature occurs at the point of neutralisation?

....

.... **(2 marks)**

Activation energy

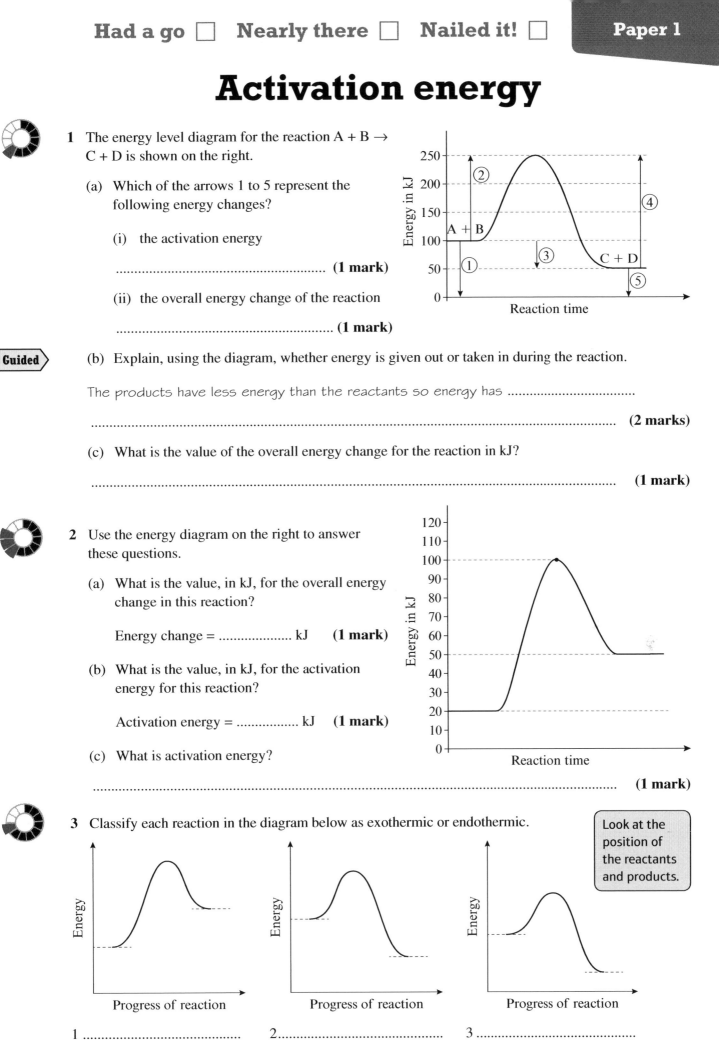

1   The energy level diagram for the reaction A + B → C + D is shown on the right.

(a) Which of the arrows 1 to 5 represent the following energy changes?

(i) the activation energy

.... **(1 mark)**

(ii) the overall energy change of the reaction

.... **(1 mark)**

> **Guided**

(b) Explain, using the diagram, whether energy is given out or taken in during the reaction.

The products have less energy than the reactants so energy has ....

.... **(2 marks)**

(c) What is the value of the overall energy change for the reaction in kJ?

.... **(1 mark)**

2   Use the energy diagram on the right to answer these questions.

(a) What is the value, in kJ, for the overall energy change in this reaction?

Energy change = ... kJ   **(1 mark)**

(b) What is the value, in kJ, for the activation energy for this reaction?

Activation energy = ... kJ   **(1 mark)**

(c) What is activation energy?

.... **(1 mark)**

3   Classify each reaction in the diagram below as exothermic or endothermic.

> Look at the position of the reactants and products.

1 ....   2 ....   3 ....

(3 marks)

Bond energies

1 Ammonia is made by the reaction of nitrogen and hydrogen as shown in the following balanced equation: $N_2(g) + 3H_2(g) \rightarrow 2NH_3(g)$

The equation can be shown using structural formulae:

N≡N + H—H H—H H—H →   N   N
                        ╱│╲   ╱│╲
                       H H H   H H H

Guided

(a) Use the bond energies in the table to calculate the energy change in this reaction.

Bond	Bond energy in kJ
N≡N	941
H—H	436
N—H	391

Energy in

1 × N≡N = 1 × 941 kJ = 941 kJ

3 × H—H = 3 × 436 kJ = ...

Total = ...

Energy change = kJ ....

Energy out

6 × N—H = 6 × ....

....

....

(3 marks)

(b) Explain whether the reaction is exothermic or endothermic in terms of bond energies.

....

.... (2 marks)

2 The balanced equation for the combustion of methane is shown below:

$CH_4(g) + 2O_2(g) \rightarrow CO_2(g) + 2H_2O$

Explain why this very exothermic reaction needs an input of energy to start it off.

....

.... (2 marks)

3 The structural formulae equation for the reaction of hydrogen and bromine is shown below:

H—H + Br—Br → 2H—Br

(a) Use the bond energies in the table to calculate the activation energy for the reaction above.

Bond	Bond energy in kJ
H—H	436
Br—Br	193

Energy change = ... kJ   (2 marks)

(b) The reaction is exothermic and the overall energy change is 103 kJ for the reaction as written.

Use this data and the information above to calculate the bond energy for the H—Br bond.

Bond energy = ... kJ   (2 marks)

Cells

1 Hydrogen fuel cells can be used to power cars and buses.

(a) Write a balanced chemical equation for the overall reaction in a hydrogen fuel cell.

.... **(2 marks)**

(b) Write the **two** half-equations for the reactions that occur at the electrodes in a hydrogen fuel cell.

....

.... **(2 marks)**

(c) Draw **one** line from each type of cell to an advantage it has.

Cell

| hydrogen fuel cell |

| rechargeable cell |

Advantage

| portable |

| not portable |

| The only waste product is water. |

| It is difficult to dispose of harmful chemicals inside it. |

(2 marks)

2 A cell was set up as shown below. Four metals, A, B, C and D, were used in turn as metal 1, and the potential difference observed was recorded in the table.

voltmeter

[0 1.80]

strip of copper metal ——

—— metal 1

—— ammonium chloride solution

Metal 1	Potential difference in V
A	+1.1
B	+2.71
C	−0.46
D	+0.58

> **Guided**

(a) Use the results in the table to place copper and the four metals, A, B, C and D, in order of reactivity from the most reactive to the least reactive. Give a reason for your answer.

If the metal is more reactive than copper then the voltage measured is positive. If the metal is less reactive than copper then the voltage measured is negative.

B, A and D are more reactive than copper because the voltage observed is positive for B, A and D, with B more reactive than ....

.... **(3 marks)**

(b) What is the potential difference if two strips of copper are used as electrodes?

.... **(1 mark)**

(c) In this experiment, the type of metal electrode was varied. What other factor can be varied in order to change the voltage of the cell?

.... **(1 mark)**

(d) Name the electrolyte in this experiment.

.... **(1 mark)**

Extended response – Energy changes

Hydrogen and chlorine react to form hydrogen chloride. Hydrogen and bromine react to form hydrogen bromide. The equations for the reactions are:

$H_2 + Cl_2 \rightarrow 2HCl$

$H_2 + Br_2 \rightarrow 2HBr$

The table below shows the bond energies relevant to these reactions.

Bond	Bond energy (kJ/mol)
H—H	436
Cl—Cl	243
H—Cl	432
H—Br	366
Br—Br	193

Calculate the energy change for each reaction and classify each reaction as exothermic or endothermic.

> Remember that breaking bonds takes in energy and making bonds gives out energy.

....

....

....

....

....

....

....

....

....

....

....

....

....

....

....

....

.... **(6 marks)**

Rate of reaction

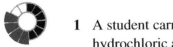

1   A student carried out an experiment to investigate the rate of reaction between marble chips and hydrochloric acid. To follow the reaction rate the student measured the mass lost by the reaction mixture with time. The results of the experiment are shown below.

Time in min	0	1	2	3	4	5	6	7	8	9	10
Mass lost in g	0.00	0.12	0.22	0.30	0.36	0.40	0.42	0.45	0.45	0.45	0.45

> **Guided**

(a)   Calculate the mean rate of reaction in g/min between:

    (i)   2 and 4 minutes $\text{Rate} = \dfrac{change}{time} = \dfrac{0.36 - 0.22}{4 - 2}$

      Rate of reaction = ... g/min   **(2 marks)**

    (ii)   4 and 6 minutes

      Rate of reaction = ... g/min   **(2 marks)**

> **Guided**

(b)   Calculate the mean rate of reaction in g/s between 1 and 3 minutes.

Time taken = 2 × 60 = 120 seconds

$\text{Rate} = \dfrac{change}{time} = \dfrac{0.3 - 0.12}{120} = \ldots\ldots\ldots\ldots\ldots\ldots$

> The time taken is 2 minutes, but here you must convert this into seconds. 1 minute = 60 seconds.

      Rate of reaction = ... g/s   **(3 marks)**

(c)   Explain when this reaction is completed.

....

.... **(2 marks)**

(d)   Calculate the mean rate of reaction up to the time the reaction is complete.

     Give your answer in g/s and to two significant figures.

> **Maths skills** Remember the first significant figure is the first non-zero number.

      Rate of reaction = ... g/s   **(4 marks)**

(e)   Draw a labelled diagram of the apparatus used to carry out this experiment.

                                 (3 marks)

Rate of reaction on a graph

1 The loss in mass was measured every minute during the reaction of magnesium and hydrochloric acid.

(a) Write a balanced chemical equation for the reaction of magnesium and hydrochloric acid.

> An acid and a metal produce a salt and hydrogen.

....

.... **(2 marks)**

(b) The table below shows the student's results.

Time in min	0	1	2	3	4	5	6	7	8	9	10
Mass lost in g	0	0.12	0.22	0.30	0.36	0.40	0.42	0.45	0.45	0.45	0.45

Plot the results from the table on the grid, and draw a line of best fit.

Mass lost in g

Time in min

> **Maths skills** Often there are 3 marks for drawing a graph: 1 mark for sensible scales, 1 mark for using at least half the grid and plotting all points and 1 mark for drawing a correct best fit line.

(3 marks)

2 The rate of reaction between calcium carbonate and hydrochloric acid was investigated by using calcium carbonate lumps and powder: $2HCl(aq) + CaCO_3(s) \rightarrow CaCl_2(aq) + H_2O(l) + CO_2(g)$

The graph on the right shows the results of measuring the volume of gas produced against time for two different experiments.

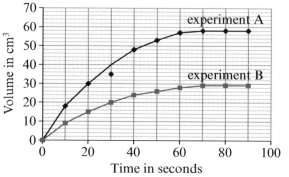

(a) Calculate the mean rate of reaction in cm^3/s between 0 and 20 seconds in experiment A. Give your answer to two significant figures.

Rate of reaction = ... cm^3/s **(2 marks)**

(b) Calculate the rate of reaction in cm^3/s between 0 and 20 seconds in experiment B. Give your answer to two significant figures.

Rate of reaction = ... cm^3/s **(2 marks)**

(c) At what time are both reactions finished?

> The reaction finishes when no more gas is produced and the graph levels off.

.... **(1 mark)**

(d) Identify any anomalous results in experiments A and B.

.... **(1 mark)**

Calculating the gradient

1   A conical flask containing
    hydrochloric acid and granulated
    zinc was placed on a balance
    and the mass recorded every
    10 seconds. The graph shows
    the results for this experiment.

    (a) Identify any anomalous
         results in this experiment.

    ....
    (1 mark)

    (b) Use the graph to calculate
         the mean rate of the reaction
         up to the time when the
         reaction is complete. Give
         your answer to two significant figures. Give the unit.

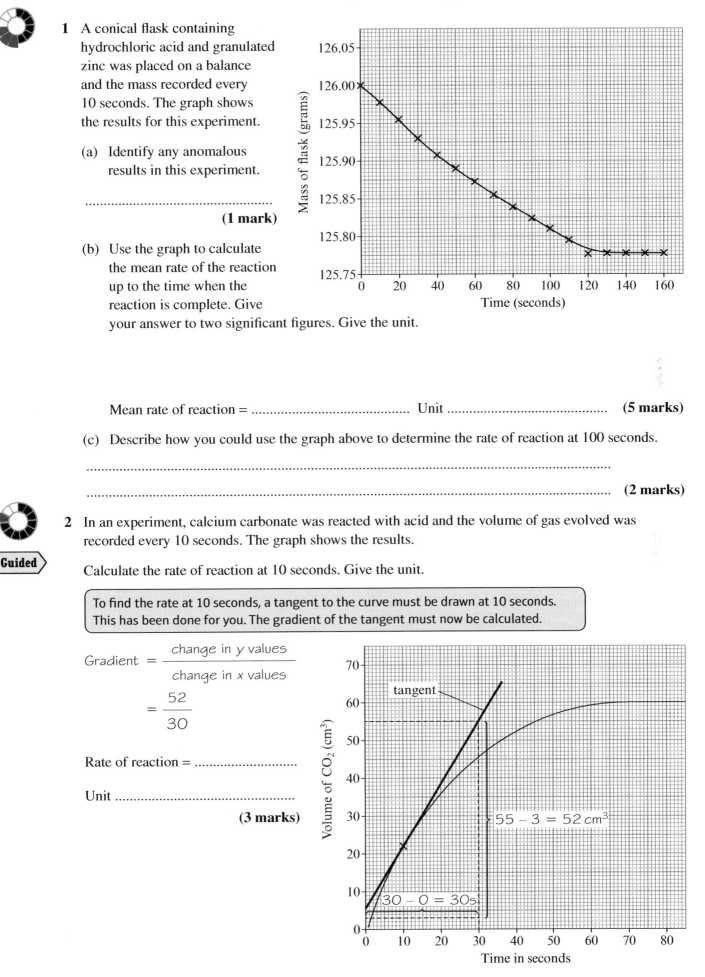

         Mean rate of reaction = ... Unit .... **(5 marks)**

    (c) Describe how you could use the graph above to determine the rate of reaction at 100 seconds.

         ....

         .... **(2 marks)**

Guided

2   In an experiment, calcium carbonate was reacted with acid and the volume of gas evolved was
    recorded every 10 seconds. The graph shows the results.

    Calculate the rate of reaction at 10 seconds. Give the unit.

    > To find the rate at 10 seconds, a tangent to the curve must be drawn at 10 seconds.
    > This has been done for you. The gradient of the tangent must now be calculated.

    Gradient = $\dfrac{\text{change in } y \text{ values}}{\text{change in } x \text{ values}}$

    $= \dfrac{52}{30}$

    Rate of reaction = ...

    Unit ....

    (3 marks)

Collision theory

1 When do chemical reactions occur?

Tick **one** box.

☐ when particles collide or touch

☐ when particles collide for a sufficient amount of time

☐ when particles collide with sufficient energy

☐ when particles mix together in a reaction vessel   **(1 mark)**

2 In an experiment, the mass lost in a reaction between calcium carbonate and hydrochloric acid changed with time as shown in the graph.

Mass loss in g (y-axis), Time in s (x-axis)

(a) Explain, in terms of particles, why the rate changes during the reaction.

....

....

....

....

....

....

.... **(4 marks)**

> **Guided**

(b) The experiment was repeated using the same mass of calcium carbonate and the same volume of hydrochloric acid, but the acid was more concentrated.

Describe and explain what would have happened to the rate of the reaction.

In the second experiment, because the acid was more concentrated, there

were .... particles in the .... volume. This means there

were more ....

.... **(4 marks)**

3 Hydrogen and bromine react to form hydrogen bromide.

> Hydrogen and bromine are both diatomic molecules.

(a) Write a balanced symbol equation for this reaction.

.... **(2 marks)**

(b) In order to react, the hydrogen and bromine molecules must have the necessary activation energy. What is activation energy?

.... **(1 mark)**

Rate: pressure, surface area

1 Which reaction will be fastest at the start of the reaction?

Tick **one** box.

☐ calcium carbonate lumps reacting with an excess of 1 mol/dm³ nitric acid

☐ calcium carbonate lumps reacting with an excess of 2 mol/dm³ nitric acid

☐ calcium carbonate powder reacting with an excess of 1 mol/dm³ nitric acid

☐ calcium carbonate powder reacting with an excess of 2 mol/dm³ nitric acid   **(1 mark)**

2 In an experiment, 0.5 g of magnesium ribbon reacted with excess dilute hydrochloric acid at room temperature. The volume of gas produced was recorded every 10 seconds. The results are shown in the graph as line B.

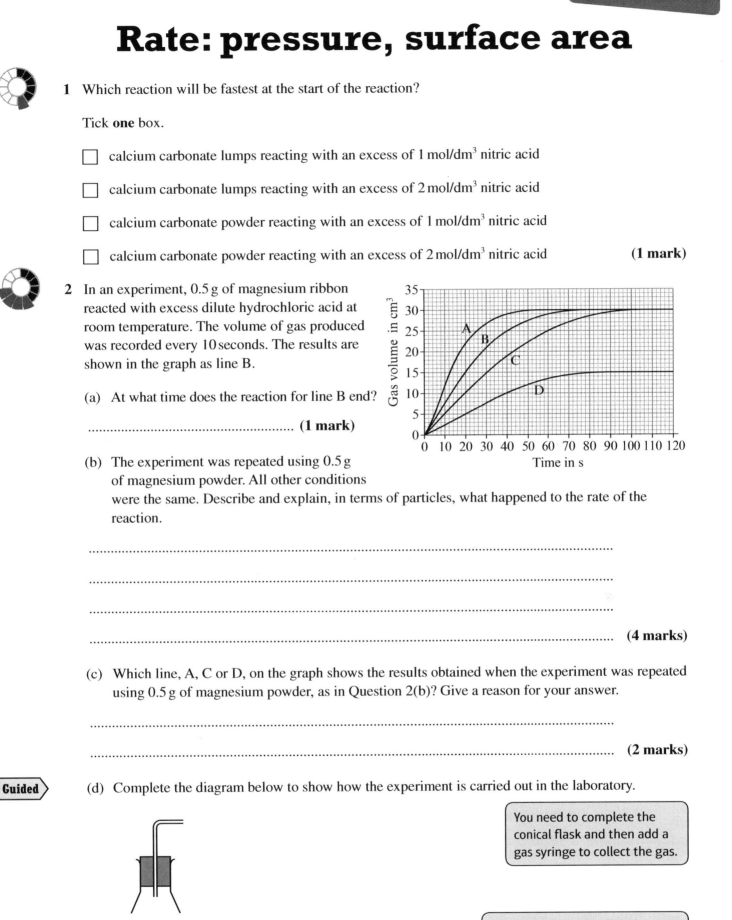

(a) At what time does the reaction for line B end?

.... **(1 mark)**

(b) The experiment was repeated using 0.5 g of magnesium powder. All other conditions were the same. Describe and explain, in terms of particles, what happened to the rate of the reaction.

....

....

....

.... **(4 marks)**

(c) Which line, A, C or D, on the graph shows the results obtained when the experiment was repeated using 0.5 g of magnesium powder, as in Question 2(b)? Give a reason for your answer.

....

.... **(2 marks)**

Guided

(d) Complete the diagram below to show how the experiment is carried out in the laboratory.

> You need to complete the conical flask and then add a gas syringe to collect the gas.

> What separate piece of apparatus is needed to record the volume of gas every 20 seconds?

(4 marks)

Rate: temperature

1   A group of students was investigating the reactions between two metals and dilute hydrochloric acid. The metals used were magnesium and zinc, and they set up the experiments as shown below. (Note: the concentration of the acid is measured in mol/dm³.)

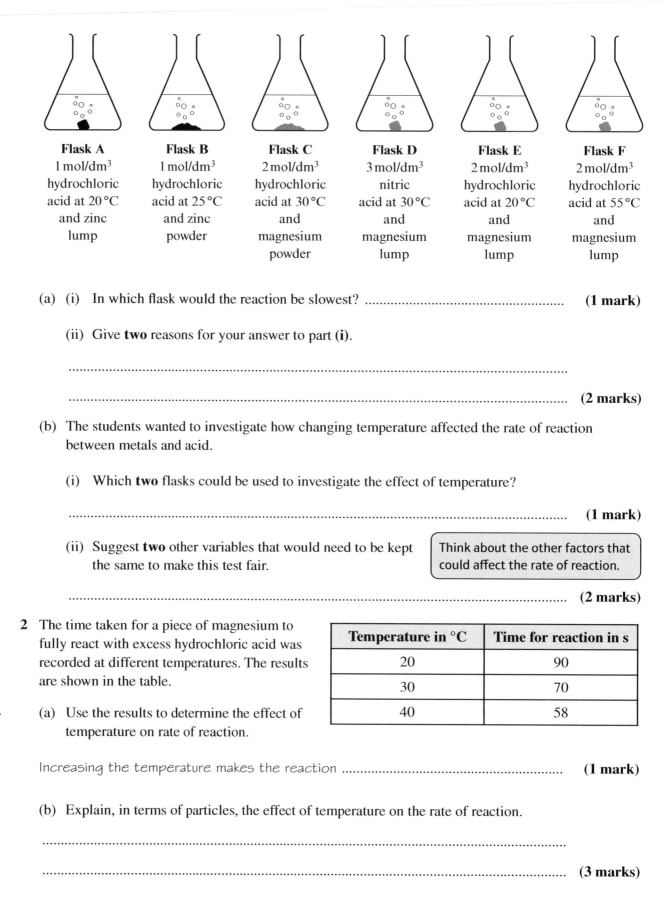

Flask A	**Flask B**	**Flask C**	**Flask D**	**Flask E**	**Flask F**
1 mol/dm³ hydrochloric acid at 20 °C and zinc lump	1 mol/dm³ hydrochloric acid at 25 °C and zinc powder	2 mol/dm³ hydrochloric acid at 30 °C and magnesium powder	3 mol/dm³ nitric acid at 30 °C and magnesium lump	2 mol/dm³ hydrochloric acid at 20 °C and magnesium lump	2 mol/dm³ hydrochloric acid at 55 °C and magnesium lump

(a) (i) In which flask would the reaction be slowest? .... **(1 mark)**

    (ii) Give **two** reasons for your answer to part (**i**).

....

.... **(2 marks)**

(b) The students wanted to investigate how changing temperature affected the rate of reaction between metals and acid.

    (i) Which **two** flasks could be used to investigate the effect of temperature?

.... **(1 mark)**

    (ii) Suggest **two** other variables that would need to be kept the same to make this test fair.

> Think about the other factors that could affect the rate of reaction.

.... **(2 marks)**

2   The time taken for a piece of magnesium to fully react with excess hydrochloric acid was recorded at different temperatures. The results are shown in the table.

Temperature in °C	**Time for reaction in s**
20	90
30	70
40	58

> Guided

(a) Use the results to determine the effect of temperature on rate of reaction.

Increasing the temperature makes the reaction .... **(1 mark)**

(b) Explain, in terms of particles, the effect of temperature on the rate of reaction.

....

.... **(3 marks)**

Practical skills

Core practical – Rate of reaction

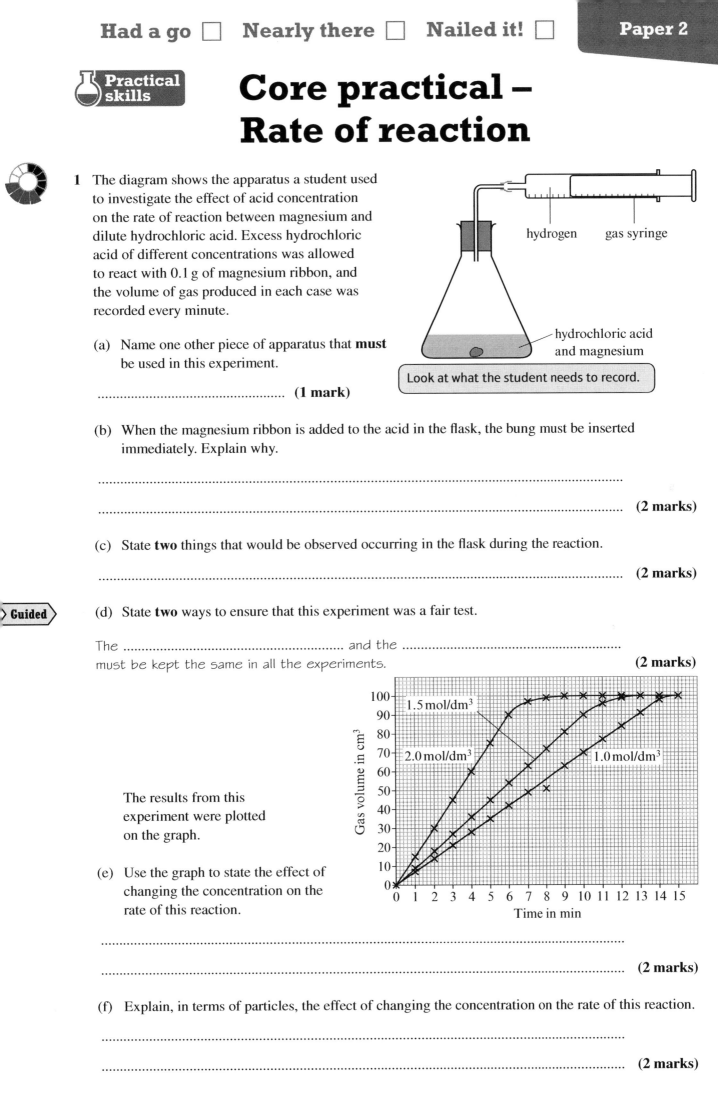

1 The diagram shows the apparatus a student used to investigate the effect of acid concentration on the rate of reaction between magnesium and dilute hydrochloric acid. Excess hydrochloric acid of different concentrations was allowed to react with 0.1 g of magnesium ribbon, and the volume of gas produced in each case was recorded every minute.

hydrogen   gas syringe

hydrochloric acid and magnesium

Look at what the student needs to record.

(a) Name one other piece of apparatus that **must** be used in this experiment.

.... **(1 mark)**

(b) When the magnesium ribbon is added to the acid in the flask, the bung must be inserted immediately. Explain why.

....

.... **(2 marks)**

(c) State **two** things that would be observed occurring in the flask during the reaction.

.... **(2 marks)**

Guided

(d) State **two** ways to ensure that this experiment was a fair test.

The .... and the ....
must be kept the same in all the experiments.   **(2 marks)**

The results from this experiment were plotted on the graph.

Gas volume in cm³

100
90
80
70
60
50
40
30
20
10
0

1.5 mol/dm³

2.0 mol/dm³   1.0 mol/dm³

0 1 2 3 4 5 6 7 8 9 10 11 12 13 14 15
Time in min

(e) Use the graph to state the effect of changing the concentration on the rate of this reaction.

....

.... **(2 marks)**

(f) Explain, in terms of particles, the effect of changing the concentration on the rate of this reaction.

....

.... **(2 marks)**

Catalysts

1   Which of the graphs A, B, C or D, shows how the mass of a catalyst changes with time during a reaction? Give a reason for your answer.

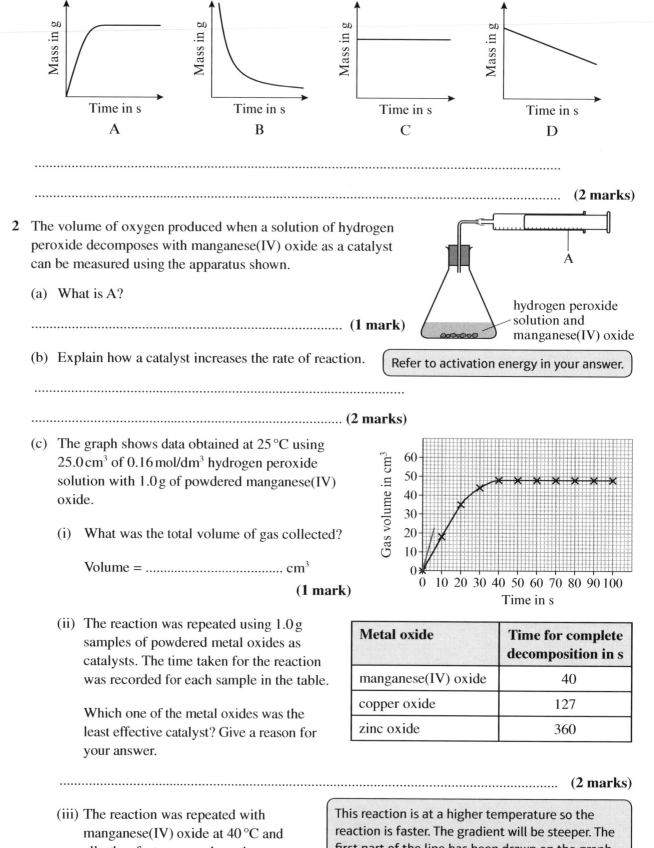

A   B   C   D

....

.... **(2 marks)**

2   The volume of oxygen produced when a solution of hydrogen peroxide decomposes with manganese(IV) oxide as a catalyst can be measured using the apparatus shown.

(a) What is A?

.... **(1 mark)**

hydrogen peroxide solution and manganese(IV) oxide

(b) Explain how a catalyst increases the rate of reaction.

| Refer to activation energy in your answer. |

....

.... **(2 marks)**

(c) The graph shows data obtained at 25 °C using 25.0 cm³ of 0.16 mol/dm³ hydrogen peroxide solution with 1.0 g of powdered manganese(IV) oxide.

(i)   What was the total volume of gas collected?

Volume = ... cm³

(1 mark)

(ii) The reaction was repeated using 1.0 g samples of powdered metal oxides as catalysts. The time taken for the reaction was recorded for each sample in the table.

Which one of the metal oxides was the least effective catalyst? Give a reason for your answer.

Metal oxide	Time for complete decomposition in s
manganese(IV) oxide	40
copper oxide	127
zinc oxide	360

.... **(2 marks)**

(iii) The reaction was repeated with manganese(IV) oxide at 40 °C and all other factors were kept the same. Sketch the line you would expect on the graph above.   **(1 mark)**

This reaction is at a higher temperature so the reaction is faster. The gradient will be steeper. The first part of the line has been drawn on the graph. Now draw the rest of the line. Remember the same volume of gas will be produced but the reaction finishes faster.

Reversible reactions

1 What is a reversible reaction?

Tick **one** box.

☐ a reaction in which heat is alternately given out and taken in as the reaction proceeds

☐ a reaction in which heat is taken in

☐ a reaction in which the products of the reaction can react to produce the original reactants

☐ a reaction in which the reactants are converted into products   **(1 mark)**

2 In the Haber process, the reaction forming ammonia from nitrogen and hydrogen can be written as shown below:

nitrogen + hydrogen $\rightleftharpoons$ ammonia

(a) Complete the balanced equation below, including state symbols.

$N_2(g)$ + ... (......) $\rightleftharpoons$ .... (......)   **(2 marks)**

(b) How does the equation show that this reaction is reversible?

.... **(1 mark)**

3 Methane can be formed when carbon monoxide reacts with hydrogen:

$CO(g) + 3H_2(g) \rightleftharpoons CH_4(g) + H_2O(g)$

> **Guided**

(a) What does the double arrow ($\rightleftharpoons$) between reactants and products mean?

This means the reaction goes .... **(1 mark)**

(b) Name the molecules that will be present when this reaction has been left for some time.

....

.... **(2 marks)**

4 The reaction between anhydrous copper sulfate and water to give hydrated copper sulfate is a reversible reaction. It is exothermic in one direction and endothermic in the opposite direction:

anhydrous copper sulfate + water $\rightleftharpoons$ hydrated copper sulfate

(a) Which direction of the reaction is the endothermic direction?

.... **(1 mark)**

(b) What is the colour change in the reaction from left to right?

.... **(1 mark)**

(c) What does hydrated mean?

.... **(1 mark)**

Equilibrium and Le Chatelier's principle

1 Hydrogen can be made by reacting methane with steam as shown in the equation below:

methane + steam → hydrogen + carbon monoxide

$$CH_4(g) + H_2O(g) \rightarrow 3H_2(g) + CO(g)$$

This reaction is reversible and an equilibrium can be reached.

(a) What does equilibrium mean?

....

.... **(2 marks)**

(b) State a necessary condition for equilibrium to occur in a reversible reaction.

.... **(1 mark)**

Guided

(c) Explain how decreasing the pressure of the reactants will affect the amount of hydrogen formed.

The position of equilibrium will move to ....

....

.... **(4 marks)**

2 The graph shows how the yield of ammonia, in the Haber process, is affected by changes to temperature and pressure.

(a) The Haber process is normally carried out at 200 atmospheres pressure. Suggest **one** advantage and **one** disadvantage of increasing the pressure in the Haber process beyond 200 atmospheres pressure.

....

....

....

Think about the effect on the yield and the cost.

.... **(2 marks)**

(b) The Haber process is normally carried out at a temperature of 450 °C.

(i) With reference to the graph, give a reason why a lower temperature of 200 °C might be an advantage.

.... **(1 mark)**

(ii) Suggest a reason why a very low temperature is not used in this process.

.... **(1 mark)**

Equilbrium: changing temperature and pressure

1   Dinitrogen tetroxide (N_2O_4) is a pale yellow gas. It decomposes in an endothermic reaction to form the brown gas nitrogen dioxide (NO_2). The equation for the reversible reaction is shown below.

$$N_2O_4(g) \rightleftharpoons 2NO_2(g)$$

pale yellow   brown

(a) State what is meant by:

(i) an endothermic reaction .... **(1 mark)**

(ii) a reversible reaction .... **(1 mark)**

(b) Explain what happens to the colour of these gases if the temperature is raised from 20 °C to 40 °C.

> Guided

If the temperature is raised, the yield in the endothermic direction will increase, so there will be more NO_2 and the colour will ....

.... **(3 marks)**

(c) Explain what happens to the colour of these gases if the gas pressure is increased.

> Look at the number of moles on each side of the equation and consider how the equilibrium position moves.

....

....

.... **(3 marks)**

2   In each of the following reversible reactions, explain how the amount of product formed at equilibrium is affected by decreasing the gas pressure.

(a) $H_2(g) + Br_2(g) \rightleftharpoons 2HBr_2(g)$
       reactants   products

....

.... **(2 marks)**

(b) $2SO_2(g) + O_2(g) \rightleftharpoons 2SO_3(g)$
       reactants   products

....

.... **(2 marks)**

3   Methanol is formed in industry by the exothermic reaction between carbon monoxide and hydrogen. The balanced equation for this reversible reaction is shown below.

$$CO(g) + 2H_2(g) \rightleftharpoons CH_3OH(g)$$

Why would the yield of methanol in this reaction be increased by lowering the temperature and raising the gas pressure?

....

.... **(2 marks)**

🧪 Practical skills Extended response – Rate of reaction and equilibria

Magnesium reacts with dilute hydrochloric acid. A student has been asked to investigate how the rate of this reaction changes when the concentration of hydrochloric acid is changed. Write a plan the student could use.

> Write a word equation for the reaction, and study it to help you decide what to measure. Make sure your plan includes how you will ensure the experiment is a fair test.

....

....

....

....

....

....

....

....

....

....

....

....

....

....

....

....

....

....

....

....

.... **(6 marks)**

> For a fair test, make sure you have controlled all the factors that might affect the rate.

Crude oil

1 The first process in oil refining separates the crude oil into fractions.

20°C ———— fuel gases C_1 to C_4

70°C

120°C ———— petrol C_5 to C_{10}

170°C ———— kerosene

230°C

heater

350°C ———— diesel oil C_{14} to C_{20}

450°C

———— residue above C_{20}

(a) Suggest a range of molecular sizes for kerosene.

.... **(1 mark)**

(b) (i) Which fraction has the highest boiling point?

.... **(1 mark)**

(ii) Which fraction will contain the molecules shown below?

| Count the carbon atoms in the structures. |

.... **(1 mark)**

(c) What is the name for the method of separation shown in the fractionating column?

.... **(1 mark)**

2 Fractional distillation produces fractions with different properties and uses. Each fraction contains similar-sized hydrocarbon molecules.

Guided

(a) Describe the changes that occur during fractional distillation.

The crude oil is heated until most of it has .... It then passes into a

.... As the mixture ....

the gases .... and condense .... **(4 marks)**

(b) The fractions can be processed to produce fuels and feedstock for the petrochemical industry.

(i) Name **two** fractions that can be processed to produce fuel.

.... **(2 marks)**

(ii) Name **three** materials produced by the petrochemical industry that are used in modern life, other than fuel.

....

.... **(3 marks)**

Alkanes

1 Crude oil is a mixture of different hydrocarbons.

(a) What is a hydrocarbon?

....

.... **(2 marks)**

(b) The table gives information about hydrocarbons called alkanes. Complete the missing information in the table.

Alkanes	Formula	Boiling point in °C
	CH_4	-162
ethane	C_2H_6	-89
propane		
butane	C_4H_{10}	0
pentane	C_5H_{12}	$+36$

(3 marks)

Guided

2 Draw the structure of:

(a) butane

$$C—C—C—C$$

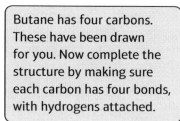

> Butane has four carbons. These have been drawn for you. Now complete the structure by making sure each carbon has four bonds, with hydrogens attached.

(1 mark)

(b) ethane

(1 mark)

(c) an alkane with five carbons.

(1 mark)

3 Which hydrocarbon is not an alkane?

> Use the general formula for alkanes to help.

Tick **one** box.

☐ CH_4 ☐ C_3H_8 ☐ C_4H_8 ☐ C_5H_{12}

(1 mark)

Properties of hydrocarbons

1 The diagram shows the apparatus used to investigate the products of combustion of hydrocarbons.

Guided

(a) Describe what the student would observe happening in tube A and tube B when the hydrocarbon had been burning for a few minutes.

> When describing observations, describe what you see happening rather than giving the names of products.

In tube A, a colourless .... would be observed.

In tube B, the limewater would change from ....

to ....

(2 marks)

(b) Name a substance that could be used to confirm the identity of the product formed in tube A and state the observations.

....

....

.... **(3 marks)**

(c) Complete the word equation for the complete combustion of the hydrocarbon.

hydrocarbon + oxygen → .... **(1 mark)**

(d) Write a balanced chemical equation for the complete combustion of butane.

.... **(2 marks)**

> Butane has four carbon atoms. Use the general formula of alkanes to work out its molecular formula.

2 Pentane (C_5H_{12}) and octane (C_8H_{18}) are both alkanes.

(a) Which of these alkanes has a higher boiling point? Give a reason for your answer.

.... **(2 marks)**

(b) Which of these alkanes is more flammable?

.... **(1 mark)**

(c) The liquid fuel used in motor car engines needs to have a low viscosity and a high flammability.

Suggest **two** reasons why these properties are important for the fuel in a car engine.

....

.... **(2 marks)**

Cracking

1 Hydrocarbons can be cracked.

(a) What does cracking mean?

.... **(2 marks)**

Guided (b) The process of cracking starts with heating the hydrocarbons to vaporise them. State two ways in which the vapours can be treated to allow cracking to occur.

Pass the vapour over a hot catalyst or ....

.... **(2 marks)**

(c) Complete the balanced symbol equation for the cracking of C_8H_{18} to form 1 mole of ethene and an alkane.

> First put in the formula of ethene on the product side and then work out the formula of the alkane. Make sure that there are the same numbers of carbon and hydrogen atoms on both sides of the equation.

.... **(2 marks)**

(d) Suggest **two** reasons why there is greater demand for the products of cracking than for C_8H_{18}.

....

.... **(2 marks)**

2 What type of reaction is cracking?

Tick **one** box.

☐ displacement   ☐ neutralisation

☐ exothermic   ☐ thermal decomposition   **(1 mark)**

3 Fractional distillation separates crude oil into fractions of similar hydrocarbons. The table below compares the fractions obtained from crude oil from three different sources.

Fraction	Crude oil A content %	Crude oil B content %	Crude oil C content %
fuel gases	6	4	9
petrol and naphtha	10	6	19
diesel and kerosene	15	10	18
fuel oil	17	20	21
bitumen and residue	52	60	33

(a) Explain which of the crude oils would have the highest viscosity.

....

.... **(2 marks)**

(b) Explain the effect that cracking would have on the percentage content of the fractions obtained from the crude oils.

....

.... **(2 marks)**

Alkenes

1   The table below shows some hydrocarbons that could be found in products from crude oil.

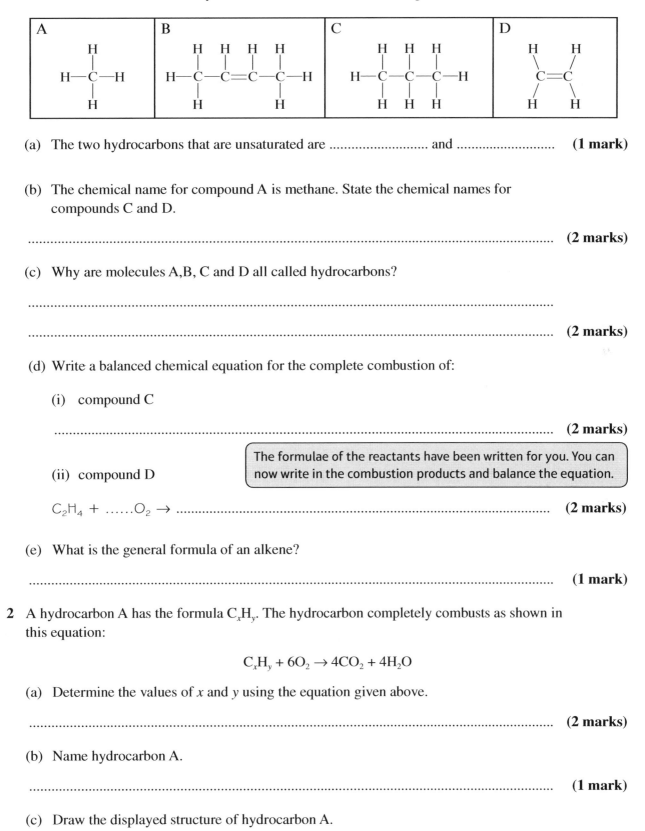

(a) The two hydrocarbons that are unsaturated are .... and .... **(1 mark)**

(b) The chemical name for compound A is methane. State the chemical names for compounds C and D.

.... **(2 marks)**

(c) Why are molecules A, B, C and D all called hydrocarbons?

....

.... **(2 marks)**

(d) Write a balanced chemical equation for the complete combustion of:

   (i) compound C

   .... **(2 marks)**

   The formulae of the reactants have been written for you. You can now write in the combustion products and balance the equation.

   (ii) compound D

   $C_2H_4 + ...O_2 \rightarrow$ .... **(2 marks)**

(e) What is the general formula of an alkene?

.... **(1 mark)**

2   A hydrocarbon A has the formula C_xH_y. The hydrocarbon completely combusts as shown in this equation:

$$C_xH_y + 6O_2 \rightarrow 4CO_2 + 4H_2O$$

(a) Determine the values of x and y using the equation given above.

.... **(2 marks)**

(b) Name hydrocarbon A.

.... **(1 mark)**

(c) Draw the displayed structure of hydrocarbon A.

(1 mark)

Reactions of alkenes

Guided

1 Propene is an alkene. It undergoes addition reactions with chlorine, bromine, hydrogen and water.

Draw displayed structural formulae for the products of the addition reactions of propene with hydrogen, water, chlorine and bromine in the boxes.

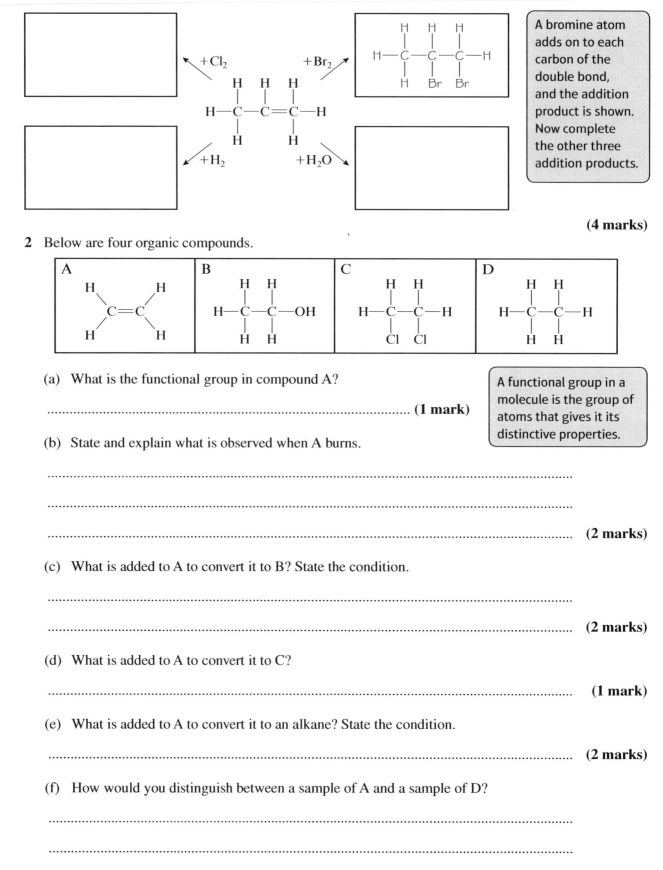

A bromine atom adds on to each carbon of the double bond, and the addition product is shown. Now complete the other three addition products.

(4 marks)

2 Below are four organic compounds.

A	B	C	D
H—C=C—H (with H above and below)	H—C—C—OH (H above, H below)	H—C—C—H (Cl below each C)	H—C—C—H (H above and below each C)

(a) What is the functional group in compound A?

.... **(1 mark)**

A functional group in a molecule is the group of atoms that gives it its distinctive properties.

(b) State and explain what is observed when A burns.

....

....

.... **(2 marks)**

(c) What is added to A to convert it to B? State the condition.

....

.... **(2 marks)**

(d) What is added to A to convert it to C?

.... **(1 mark)**

(e) What is added to A to convert it to an alkane? State the condition.

.... **(2 marks)**

(f) How would you distinguish between a sample of A and a sample of D?

....

....

.... **(3 marks)**

Alcohols

1 The fourth member of the alcohol series is called butanol.

    (a) What is the functional group in the butanol molecule? .... **(1 mark)**

    (b) What is the molecular formula of butanol? .... **(1 mark)**

2 (a) What **two** products would be formed by the complete combustion of butanol?

.... **(2 marks)**

    (b) Suggest **two** possible uses for butanol.

    Butanol could be used as a solvent or as a .... **(2 marks)**

    (c) Write a balanced chemical equation for the combustion of butanol.

.... **(2 marks)**

3 Methanol and ethanol are two members of the alcohol series. They have similar molecular structures and similar chemical properties.

    (a) Write down the molecular formula for methanol and the displayed structural formula for ethanol.

    ....

    (2 marks)

    (b) Draw a circle around the correct answer to complete the following sentences about the properties of the alcohols methanol and ethanol.

    Both these alcohols react with

| magnesium |
| copper |
| sodium |

metal to produce

| oxygen |
| hydrogen |
| carbon dioxide |

gas.

    (2 marks)

4 Which structure below shows propanol?

Tick **one** box.

☐   ☐   ☐   ☐

    (1 mark)

5 Ethanol is produced by fermentation of sugars.

Describe the conditions needed for fermentation.

....

....

.... **(3 marks)**

END of C 10

Carboxylic acids

1 The structural formula of butanoic acid, the fourth member of a homologous series of carboxylic acids, is shown on the right.

(a) What is the molecular formula of butanoic acid?

.... **(1 mark)**

(b) Draw a circle around the functional group of butanoic acid. **(1 mark)**

2 (a) Name the alcohol that could be oxidised to make butanoic acid. .... **(1 mark)**

(b) The third member of the carboxylic acid series is called propanoic acid. Write down the molecular formula of propanoic acid and draw its displayed structural formula.

....

(2 marks)

(c) What gas is produced when carboxylic acids, like propanoic acid and butanoic acid, react with metal carbonates?

.... **(1 mark)**

> **Guided**

(d) Describe another common reaction of carboxylic acids, naming the general reactant and product involved.

Carboxylic acids react with alcohols, forming .... **(2 marks)**

3 The table below shows some information about two acid solutions.

	Hydrochloric acid	Ethanoic acid
Formula	HCl	
Concentration	$0.1 \, mol/dm^3$	$0.1 \, mol/dm^3$
pH	1	
Time taken to dissolve 1 g of CaCO$_3$		6 minutes

(a) Complete the missing information in the table. **(3 marks)**

(b) The equation below represents what happens when ethanoic acid dissolves in water.

$$CH_3COOH(aq) \rightleftharpoons CH_3COO^-(aq) + H^+(aq)$$

Use this equation to describe the difference between a weak acid like ethanoic acid and a strong acid like hydrochloric acid.

....

.... **(2 marks)**

(c) Explain the difference in pH between $0.1 \, mol/dm^3$ hydrochloric acid and $0.1 \, mol/dm^3$ ethanoic acid.

....

....

.... **(3 marks)**

Addition polymers

1 Most polymers are made from molecules obtained by the refining of crude oil.

   (a) The diagram below shows the formation of one particular polymer. Write the names of the
        reactant and product in this reaction.

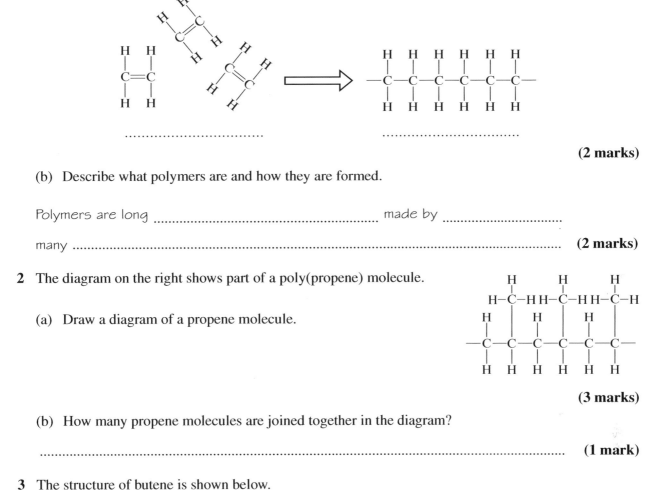

   ....

                                                                        (2 marks)

   (b) Describe what polymers are and how they are formed.

   Polymers are long .... made by ....

   many .... **(2 marks)**

2 The diagram on the right shows part of a poly(propene) molecule.

   (a) Draw a diagram of a propene molecule.

                                                                        (3 marks)

   (b) How many propene molecules are joined together in the diagram?

   .... **(1 mark)**

3 The structure of butene is shown below.

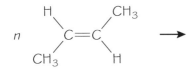

   Draw the structure to show the formation of the polymer from
   the butene monomer.

   The *n* shows there are many butene
   molecules. On the right, draw the
   monomer again, but without the
   double bond. Draw a long bond on
   each side of the C=C, then draw
   brackets through the long bonds
   and write in *n* after the brackets.

                                                                        (2 marks)

4 Draw the monomer that forms the polymer shown below.

   Remember that a
   monomer must have
   a C=C double bond.

                                                                        (1 mark)

Condensation polymers

1 Which of the options below is a condensation polymer?

Tick **one** box.

☐ graphite ☐ poly(ethene)

☐ polyester ☐ poly(propene) **(1 mark)**

2 There are two types of polymerisation, condensation and addition.

(a) What is condensation polymerisation?

....

.... **(2 marks)**

> **Guided**

(b) Describe **two** differences between condensation and addition polymerisation.

> Think about the number of products formed in each type of polymerisation.

Condensation polymers are formed from monomers with two functional groups, but
addition polymers are formed from monomers ....

....

.... **(2 marks)**

(c) Draw the structure of the two monomers that could be used to make the polymer
shown below.

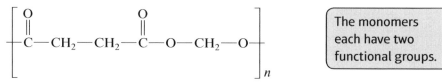

> The monomers
> each have two
> functional groups.

(2 marks)

3 Draw the structure of the condensation polymer formed from the monomers shown below.

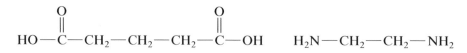

(2 marks)

Biochemistry

1 Four different organic molecules are shown in the diagram below.

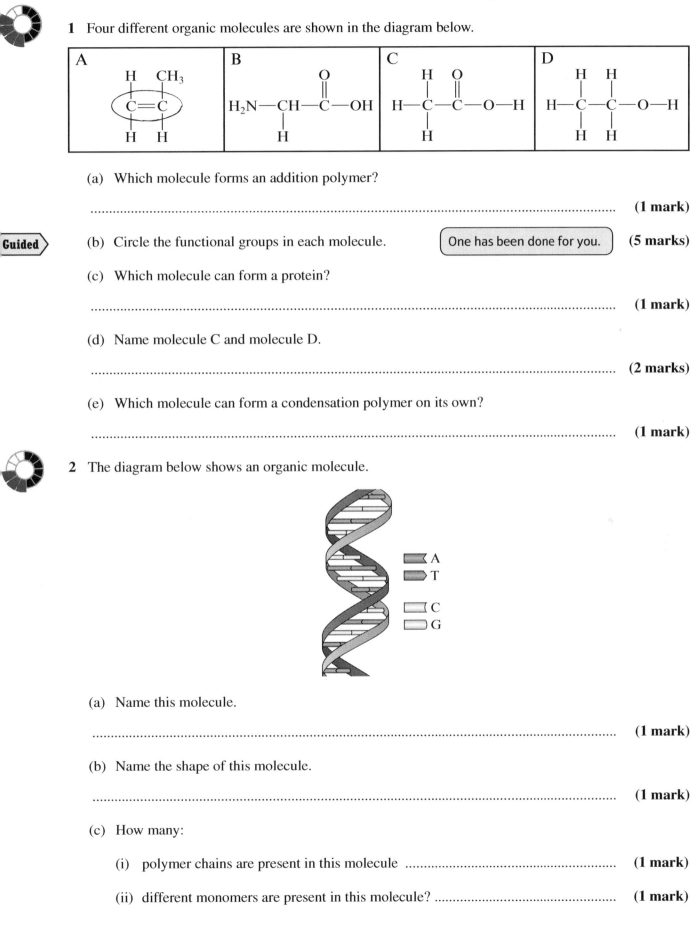

(a) Which molecule forms an addition polymer?

.... **(1 mark)**

(b) Circle the functional groups in each molecule.   [One has been done for you.]   **(5 marks)**

(c) Which molecule can form a protein?

.... **(1 mark)**

(d) Name molecule C and molecule D.

.... **(2 marks)**

(e) Which molecule can form a condensation polymer on its own?

.... **(1 mark)**

2 The diagram below shows an organic molecule.

(a) Name this molecule.

.... **(1 mark)**

(b) Name the shape of this molecule.

.... **(1 mark)**

(c) How many:

(i) polymer chains are present in this molecule .... **(1 mark)**

(ii) different monomers are present in this molecule? .... **(1 mark)**

Extended response – Organic chemistry

A dilute solution of ethanol can be produced in the laboratory by fermentation. It can then be concentrated by distillation, as shown in the diagram below. Describe the processes of fermentation and distillation.

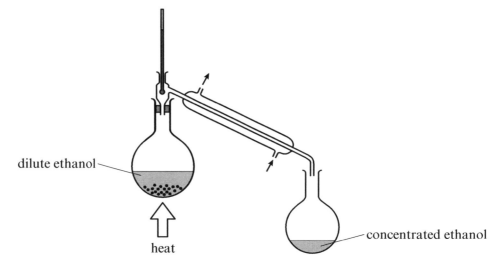

dilute ethanol

heat

concentrated ethanol

> When describing fermentation you need to state the starting materials. The question also asks about distillation, so you should describe how the apparatus shown in the diagram above can be used to concentrate the ethanol.

....

....

....

....

....

....

....

....

....

....

....

....

....

.... **(6 marks)**

Pure substances and formulations

1 A solid is thought to be pure aspirin. What is the best way to test its purity?

Tick **one** box.

☐ Determine the density.   ☐ Determine the melting point.

☐ Determine the pH.   ☐ Determine the flame colour.   **(1 mark)**

2 Iron and steel are both used in buildings. Steel is an alloy of iron and carbon.

(a) Explain why steel is a formulation.

....

....

.... **(2 marks)**

Guided

(b) Name **two** other examples of formulations.

Medicines and .... **(2 marks)**

(c) The table below shows some data about elements and alloys.

Substance	Melting point in °C	Boiling point in °C
A	420	913
B	1420–1536	2535–2545
C	−33	355

If the melting point is above 20°C, then the substance is a solid at room temperature.

(i) Classify A, B and C in the table as solids, liquids or gases at room temperature (20°C).

....

....

.... **(3 marks)**

(ii) Classify A, B and C in the table as elements or formulations. Give reasons for your answers.

....

....

.... **(3 marks)**

3 The melting point of a substance was recorded. How can this melting point be used to

(a) identify the substance

.... **(1 mark)**

(b) determine whether the substance is pure?

.... **(1 mark)**

Practical skills

Core practical – Chromatography

Guided

1 A student used the apparatus shown in the diagram to separate the substances mixed together in some purple food dye.

(a) Describe each part of the apparatus set-up.

A ....

B ....

C beaker

D ....

(4 marks)

(b) Describe what is wrong with the set-up and give a reason why it will not work as shown.

....

.... **(2 marks)**

2 A student investigated an orange drink in the laboratory using chromatography to determine whether the drink contained dyes X, Y and Z. The results are shown in the diagram on the right.

solvent front

start line drawn in pencil

X   Y   Z   drink

(a) How many dyes were in the orange drink? .... **(1 mark)**

(b) Why is the start line drawn in pencil?

.... **(1 mark)**

(c) Is dye Z a pure substance? Give a reason for your answer.

.... **(1 mark)**

(d) Calculate the R_f value of dye X.

> You need to use a ruler to help you.

(2 marks)

(e) Explain how paper chromatography separates mixtures of substances.

....

....

.... **(3 marks)**

Practical skills

Tests for gases

1 A burning splint is lowered into a gas jar. Which gas, if present in the jar, will allow the splint to burn vigorously?

Tick **one** box.

☐ carbon dioxide

☐ helium

☐ neon

☐ oxygen   **(1 mark)**

2 Calcium carbonate and hydrochloric acid were reacted in a test tube and the gas produced was bubbled into limewater.

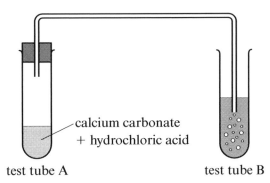

calcium carbonate + hydrochloric acid

test tube A   test tube B

(a) Write a balanced chemical equation for the reaction of calcium carbonate with hydrochloric acid.

.... **(2 marks)**

(b) State and explain what was observed in test tube B.

....

....

| What gas is produced when an acid and a carbonate react? |

.... **(3 marks)**

Guided (c) Write the chemical name and formula for limewater.

Name   *calcium hydroxide*

Formula .... **(2 marks)**

3 Describe how you would identify four gases: hydrogen, chlorine, helium and oxygen.

....

....

....

....

.... **(4 marks)**

 **Practical skills** **Tests for cations**

1 Which solution will give a coloured precipitate when a few drops of sodium hydroxide solution are added?

Tick **one** box.

☐ aluminium sulfate solution ☐ iron(II) sulfate solution

☐ calcium nitrate solution ☐ magnesium nitrate solution **(1 mark)**

2 A forensic scientist working at a crime scene tested an unknown solid found on a suspect's shoe. The results of their observations are shown below.

Appearance	Solubility	Test 1: Flame test	Test 2: Effect of adding dilute hydrochloric acid
white solid	insoluble	red flame	gas bubbles formed that turn limewater cloudy

You need to learn the flame colours produced by the following ions: lithium, sodium, potassium, calcium and copper.

(a) Describe how to carry out a flame test.

....

.... **(2 marks)**

Guided (b) Why are at least two tests needed to identify any ionic substance?

Ionic substances contain two ions and... **(2 marks)**

(c) Name the gas that is produced in test 2. .... **(1 mark)**

(d) Suggest a possible name for the unknown white solid. .... **(2 marks)**

3 The effect of adding sodium hydroxide solution to solutions of different metal ions is shown below.

Cation	Symbol	Effect of adding sodium hydroxide solution
aluminium	$Al^{3+}(aq)$	white solid formed
magnesium	$Mg^{2+}(aq)$	white solid formed
copper(II)		blue solid formed
iron(II)	$Fe^{2+}(aq)$	
iron(III)	$Fe^{3+}(aq)$	

(a) Name the type of reaction that occurs in these tests. .... **(1 mark)**

(b) Complete the missing information in the table. **(3 marks)**

(c) Describe how this test can be used to tell the difference between solutions containing aluminium and magnesium ions.

....

.... **(2 marks)**

(d) Complete the balanced equation, with state symbols, for the reaction that occurs when sodium hydroxide solution is added to calcium nitrate solution.

$2NaOH(aq) + Ca(NO_3)_2(aq) \rightarrow$ .... (......) $+ 2NaNO_3$ (......) **(3 marks)**

Practical skills

Tests for anions

1 A group of students was given a white soluble solid that was thought to be either potassium carbonate or potassium sulfate.

Describe how the students could test the white solid to see if it contained carbonate ions or sulfate ions.

First add hydrochloric acid. If carbonate ions are present a gas is produced that will turn

....

To test for sulfate ions, add ....

.... **(2 marks)**

2 Describe how to carry out **one** test for each pair of substances to identify which is which. Use chemicals from the list in the box.

barium chloride solid   distilled water   hydrochloric acid solution
nitric acid solution   silver nitrate solid   sodium hydroxide solution

(a) sodium bromide and sodium chloride

....

....

.... **(3 marks)**

(b) sodium sulfate and sodium chloride

....

....

.... **(3 marks)**

3 A sample of river water, taken downstream from a large town, was tested for pollution by analytical chemists working for the water authority.

The water sample was first evaporated until 5% was left and then tested as follows:

Test 1: Flame test results	Test 2: Adding sodium hydroxide solution	Test 3: Adding dilute hydrochloric acid	Test 4: Adding acidified silver nitrate solution
no colour produced	white precipitate formed, which dissolved when excess hydroxide was added	no effect	yellow solid formed

(a) Suggest what tests 1 and 2 would tell the analytical chemists about the river water.

....

.... **(2 marks)**

(b) Suggest what tests 3 and 4 would tell the analytical chemists about the river water.

....

.... **(2 marks)**

Flame emission spectroscopy

1 Which method is best for analysing a solution in order to determine if it contains both magnesium ions and calcium ions?

Tick **one** box.

> Do both ions have a characteristic flame test colour?

☐ Add a few drops of sodium hydroxide solution.

☐ Add a few drops of nitric acid solution.

☐ Carry out flame emission spectroscopy.

☐ Carry out a flame test. **(1 mark)**

2 The diagram below gives the flame emission spectra of four metal ions and of one mixture of two metal ions.

Li⁺

Na⁺

Ca²⁺

Cu²⁺

Mixture

(a) Use the spectra to identify the two metal ions in the mixture.

.... **(2 marks)**

(b) Explain whether the two metal ions in the mixture could be identified by using a flame test, or by adding sodium hydroxide solution.

....

....

.... **(2 marks)**

(c) Give **two** advantages of using flame emission spectroscopy, rather than chemical tests, to identify the ions in a mixture.

....

.... **(2 marks)**

(d) Write the electronic configuration of a potassium ion.

> The potassium atom has 19 electrons. First write the electronic configuration of the atom. Then, remember that the potassium ion is K⁺, so it has lost one electron.

.... **(1 mark)**

Practical skills

Core practical – Identifying a compound

Guided

1 Solid A is thought to be an aluminium or magnesium halide. Describe experiments that you could carry out to determine if solid A contains aluminium or magnesium ions and identify the halide ion present.

> Include practical details in your description.

Dissolve the sample in deionised water. Add a few drops of sodium hydroxide solution

....

....

....

....

.... **(5 marks)**

2 A mixture of two ionic compounds was analysed to determine the ions present in the mixture. The two ionic compounds have the same anion.

The results of the tests are given in the table below.

Description of test	Observations
Test 1 Flame test	yellow flame
Test 2 A sample of the mixture was dissolved in deionised water and sodium hydroxide solution was added.	white precipitate, which dissolves in excess sodium hydroxide solution
Test 3 A sample of the mixture was dissolved in deionised water, and nitric acid and drops of silver nitrate solution were added.	no effervescence, yellow precipitate

Use the information in the table to answer the questions below.

(a) Using the evidence from test 1, name **one** cation present in the mixture.

.... **(1 mark)**

(b) Using the evidence from test 2, name and give the formula for the other cation present.

.... **(2 marks)**

(c) Using the evidence from test 3, write the formula of the anion present in the mixture.

.... **(1 mark)**

(d) Suggest the names of **two** compounds present in the mixture.

.... **(2 marks)**

(e) Write an ionic equation with state symbols for the formation of the precipitate in test 3.

.... **(2 marks)**

Extended response – Chemical analysis

Plan an experiment to positively identify each ion in unlabelled samples of each of the following solutions:

- magnesium sulfate solution
- sodium chloride solution
- iron(II) iodide solution
- magnesium bromide solution.

> What cation tests do you need to use?
> What anion tests do you need to use?

....

....

....

....

....

....

....

....

....

....

....

....

....

....

....

....

....

....

....

....

.... **(6 marks)**

The early atmosphere and today's atmosphere

1 Which unreactive gas makes up most of the Earth's atmosphere today?

Tick **one** box.

☐ carbon dioxide   ☐ nitrogen

☐ helium   ☐ oxygen   **(1 mark)**

2 The proportions of the main gases in our atmosphere have not changed much over the past 200 million years.

  (a) Complete the table, to show the percentages of the **two** main gases in the Earth's atmosphere.

Main gas	% in atmosphere
nitrogen	

(2 marks)

  (b) The atmosphere also contains small amounts of other gases, for example argon, water vapour, carbon dioxide and hydrogen.

     Which of these other gases is a noble gas? .... **(1 mark)**

  (c) Name **two** gases that were present in the early atmosphere, which are not present in today's atmosphere.

.... **(2 marks)**

3 A group of students burned some magnesium in air. The volume of air reduced as the magnesium reacted with the oxygen in the air. The students recorded the volumes of air and the temperature.

The results of their experiment are shown below.

	Air volume in cm³	Temperature in °C
Experiment start	200	20
Experiment end	172	20

  (a) Why is it important that all measurements of gas volumes are made at the same temperature?

.... **(1 mark)**

  (b) Complete the balanced symbol equation for the reaction of magnesium with oxygen.

     .... Mg + ... $\rightarrow$ .... MgO   **(2 marks)**

  (c) From the results obtained by the students, calculate the percentage of oxygen in the sample of air.

....

.... **(2 marks)**

  (d) Suggest a possible reason for the percentage of oxygen calculated by this experiment being less than the actual value.

....

.... **(2 marks)**

Evolution of the atmosphere

1 The table below shows the main gases in the Earth's atmosphere today and 3.5 billion years ago.

Earth's atmosphere today	Early Earth's atmosphere (3.5 billion years ago)
nitrogen 78%	carbon dioxide 95.5%
oxygen 21%	nitrogen 3.1%
argon 0.9%	argon 1.2%
carbon dioxide 0.04%	methane 0.2%

Guided

(a) Compare the composition of gases in the Earth's early atmosphere with the atmosphere today.

The early atmosphere contained no oxygen, whereas ....

....

....

.... **(3 marks)**

(b) Explain why the data on the Earth's atmosphere today will be more accurate than the data on the early Earth's atmosphere.

....

.... **(1 mark)**

(c) Scientists think that the atmosphere has changed due to the presence of plants and algae on the Earth. Explain how the presence of algae and plants could change the atmosphere.

> Think about the chemical reactions in plants that use gases from the air for life processes.

....

.... **(2 marks)**

2 As the Earth cooled, water vapour in the atmosphere condensed, and the seas and oceans formed. Along with other changes, this event had a major effect on the make-up of the atmosphere.

(a) When carbon dioxide dissolves in water, carbonic acid is formed. Complete the balanced symbol equation for this chemical reaction, including state symbols.

$H_2O(l)$ + ... $\rightarrow H_2CO_3(aq)$ **(1 mark)**

(b) The dissolved carbon dioxide in the seas and oceans can be a problem for marine life. However, some sea creatures need the carbon dioxide for growth.

(i) What do some marine animals make with the dissolved carbon dioxide?

.... **(1 mark)**

(ii) What kind of sedimentary rocks do they eventually form?

.... **(1 mark)**

(c) Algae and plants produced the oxygen that is now in the atmosphere by photosynthesis. Balance the equation for this reaction.

......CO_2 + ...H_2O $\rightarrow$ ......$C_6H_{12}O_6$ + ...O_2 **(1 mark)**

Greenhouse gases

1   The graph below shows how the percentage of carbon dioxide in the atmosphere has changed over the past 4500 million years.

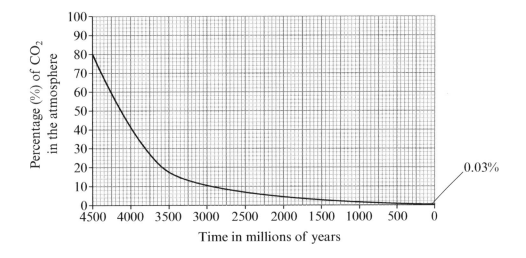

Guided   (a)   Carbon dioxide is a greenhouse gas. Name, and give the formula or symbol for, **two** other greenhouse gases.

Water vapour, which has formula .... , and ....

.... **(4 marks)**

(b)   Describe how carbon dioxide helps to maintain temperatures on Earth.

....

....

....

....

.... **(3 marks)**

(c)   State **two** conclusions that can be drawn from the diagram above.

....

....

.... **(2 marks)**

(d)   It is thought that the percentage of carbon dioxide in the atmosphere has changed in the last 100 years. State **two** human activities that may have contributed to this change.

....

.... **(2 marks)**

Global climate change

1 Due to human activity the levels of carbon dioxide in our atmosphere have been increasing over the last 100 years.

Guided

(a) Explain why destroying large areas of forest causes increased levels of carbon dioxide in the atmosphere.

During photosynthesis, plants take in ....

.... **(2 marks)**

(b) Describe how **one** other human activity is thought to be responsible for increasing carbon dioxide levels in the atmosphere.

....

.... **(2 marks)**

(c) Describe **two** environmental problems caused by the increased levels of carbon dioxide.

....

.... **(2 marks)**

2 The graph shows the changes in average world temperatures and the carbon dioxide levels over the past 400 000 years.

(a) Describe the relationship between carbon dioxide levels and average world temperatures shown by the graph.

....

....

....

.... **(2 marks)**

Vostok (Antarctica) ice core records

Temperature change in °C

CO₂ concentration in ppm

400 350 300 250 200 150 100 50 0
Time before present in thousands of years

(b) Why are scientists so worried about the trends these graphs show over recent years?

....

.... **(2 marks)**

3 The average world temperatures have been increasing in recent years. If this continues it may cause drastic changes to our environment that will affect our way of life.

(a) Define global warming.

.... **(1 mark)**

(b) Suggest **two** possible environmental changes that could be brought about by increasing average world temperatures.

....

.... **(2 marks)**

Carbon footprint

1 What is meant by the term carbon footprint?

Tick **one** box.

☐ the amount of carbon in a substance

☐ the total amount of carbon dioxide emitted over the full life cycle of a substance

☐ the total amount of all greenhouse gases emitted over the full life cycle of a substance

☐ the percentage of carbon dioxide formed from burning a substance   **(1 mark)**

2 The graph shows some factors that contribute to the carbon footprint of an average person living in the UK.

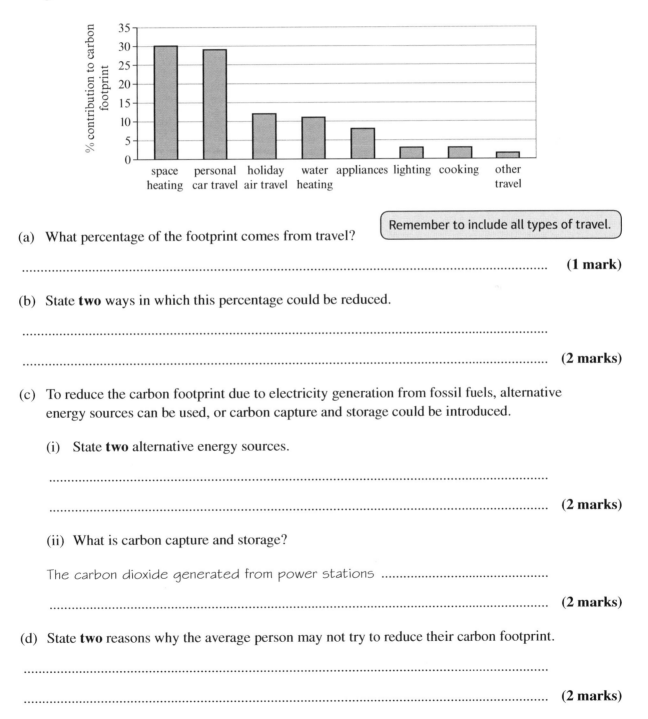

(a) What percentage of the footprint comes from travel?

> Remember to include all types of travel.

.... **(1 mark)**

(b) State **two** ways in which this percentage could be reduced.

....

.... **(2 marks)**

(c) To reduce the carbon footprint due to electricity generation from fossil fuels, alternative energy sources can be used, or carbon capture and storage could be introduced.

    (i) State **two** alternative energy sources.

....

.... **(2 marks)**

Guided   (ii) What is carbon capture and storage?

    The carbon dioxide generated from power stations ....

.... **(2 marks)**

(d) State **two** reasons why the average person may not try to reduce their carbon footprint.

....

.... **(2 marks)**

Atmospheric pollution

1 Burning fuels that contain carbon can produce carbon dioxide, carbon monoxide and soot (carbon).

(a) Which of these products are **not** formed by complete combustion of the fuel?

.... **(1 mark)**

(b) All three of the products can cause different environmental problems. Describe **one** problem caused by each of the three products.

....

....

.... **(3 marks)**

2 The flue gases from a coal-fired power station chimney were analysed. The results are shown in the table on the right.

(a) Complete the table to describe where each of these gases has come from.

Flue gas	Abundance in %	Source of gas
nitrogen	66	
carbon dioxide	18	from burning carbon in the fuel
oxygen	10	from air
sulfur dioxide	4	

(2 marks)

(b) What is the percentage of other gases present in the flue gases?

> **Maths skills** Remember that % means out of 100.

.... **(1 mark)**

(c) Sulfur dioxide is often removed from flue gases by a chemical process. Explain why this process is necessary.

....

.... **(2 marks)**

3 When petrol burns in an engine, several pollutants are formed. Describe how each of the pollutants below is formed.

> **Guided**

(a) sulfur dioxide

When the petrol burns, the sulfur oxidises and sulfur dioxide is produced. **(1 mark)**

(b) nitrogen oxides

.... **(1 mark)**

(c) carbon monoxide

.... **(1 mark)**

(d) soot

.... **(1 mark)**

Extended response – Using resources

Ammonia is made from the reaction of nitrogen with hydrogen in the Haber process:

$$N_2 + 2H_3 \rightleftharpoons 2NH_3$$

The forward reaction is exothermic. Ammonia production is carried out at a temperature of 450 °C and at a pressure of 200 atmospheres with an iron catalyst.

Explain, in terms of equilibrium principles, why each condition is used and the compromises made in using those conditions.

> To think about the effect of pressure, it is useful to write down the number of molecules on each side of the equation. The reaction is exothermic, so think of the effect of high and low temperature on the equilibrium.

....

....

....

....

....

....

....

....

....

....

....

....

....

....

....

....

....

....

.... **(6 marks)**

Practice Paper 1

Time allowed: 1 hour 45 minutes

Total marks: 100

AQA publishes official Sample Assessment Material on its website. This practice exam paper has been written to help you practise what you have learned and may not be representative of a real exam paper.

1 Some substances can be classified as elements, compounds or mixtures.

   (a) Which substance is a compound?

   ☐ brass

   ☐ chlorine

   ☐ methane

   ☐ sodium chloride solution **(1 mark)**

   (b) The circles in Figure 1 show atoms.

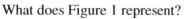

Figure 1

   What does Figure 1 represent?

   ☐ a mixture of an element and two compounds

   ☐ a mixture of a compound and two elements

   ☐ a mixture of three different elements

   ☐ a mixture of three different compounds **(1 mark)**

   (c) What is the best method of separating sand from a mixture of sand and seawater?

   ☐ chromatography

   ☐ crystallisation

   ☐ filtration

   ☐ fractional distillation **(1 mark)**

   (Total for Question 1 is 3 marks)

2 Figure 2 shows the outer electrons in an atom of magnesium and an atom of chlorine.

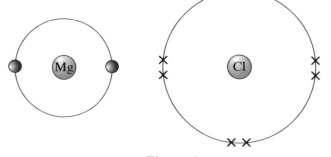

Figure 2

   (a) Magnesium and chlorine react to form the ionic compound magnesium chloride.

   (i) Name the type of bonding in the element magnesium. **(1 mark)**

   (ii) Name the type of bonding in the element chlorine. **(1 mark)**

(b) Describe what happens when an atom of magnesium reacts with two atoms of chlorine. Give your answer in terms of electron transfer. Give the formulae of the ions formed. **(5 marks)**

(c) Chlorine can also form covalent bonds. Complete the dot-and-cross diagram in Figure 3 to show the covalent bonding in a molecule of hydrogen chloride. Show the outer shell electrons only. **(2 marks)**

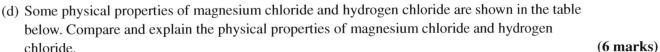

Figure 3

(d) Some physical properties of magnesium chloride and hydrogen chloride are shown in the table below. Compare and explain the physical properties of magnesium chloride and hydrogen chloride. **(6 marks)**

	Magnesium chloride	**Hydrogen chloride**
Melting point in °C	714	−114
Electrical conductivity	conducts when molten or dissolved	does not conduct

(e) Hydrogen chloride dissolves in water and produces H^+ and Cl^- ions.

Name the ions produced. **(2 marks)**

(Total for Question 2 is 17 marks)

3 There are many different structural models of an atom.

(a) Describe the differences between the plum pudding model of an atom and the nuclear model of the atom we use today. **(4 marks)**

(b) The table below gives some information about four different particles, A, B, C and D. Some particles are **atoms** and some are **ions**. The letters A, B, C and D are not chemical symbols.

(i) Complete the table.

Particle	Atomic number	Mass number	Number of protons	Number of neutrons	Number of electrons	Electronic structure
A		23	11			2,8,1
B		27	13			2,8
C			20	20	20	
D	7	14			10	

(4 marks)

(ii) Particle A has the electronic structure 2,8,1. What does this tell you about the position of A in the periodic table? **(2 marks)**

(iii) Particle D is a negative ion. What is the charge on this ion? **(1 mark)**

(c) What is the approximate radius of an atom?

☐ 1×10^{-1} m

☐ 1×10^{-8} m

☐ 1×10^{-10} m

☐ 1×10^{-14} m **(1 mark)**

(Total for Question 3 is 12 marks)

4 An experiment was carried out to determine if the reaction between hydrochloric acid and sodium hydroxide was exothermic.

This is the method used.

1. 25.0 cm³ of 0.10 mol/dm³ hydrochloric acid was measured out and placed in a polystyrene cup.

2. The temperature of the hydrochloric acid was recorded.

3. 25.0 cm³ of sodium hydroxide solution was then added gradually in 5.0 cm³ portions to the hydrochloric acid, stirring after each addition.

4. The temperature of the reaction mixture was recorded.

The table below shows the results.

Volume of sodium hydroxide added in cm³	0.0	5.0	10.0	15.0	20.0	25.0
Temperature of reaction mixture in °C	20.5	21.5	22.5	23.5	25.2	28.0

(a) Plot a graph of the results. Draw a line of best fit on Figure 4.

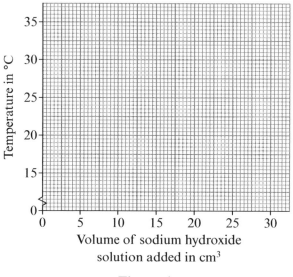

Figure 4

(3 marks)

(b) How does your graph show that this reaction was exothermic? **(1 mark)**

(c) What piece of apparatus could be used to add the sodium hydroxide solution to the acid? **(1 mark)**

(d) Suggest one improvement that could be made to the apparatus used that would give more accurate results. Give a reason for your answer. **(2 marks)**

(e) Write a balanced chemical equation for the reaction between sodium hydroxide and hydrochloric acid. **(2 marks)**

(f) Calculate the number of moles of hydrochloric acid placed in the polystyrene cup. **(2 marks)**

(g) Hydrochloric acid is a strong acid. What is meant by strong acid? **(2 marks)**

(h) A student wished to change the experiment to determine the temperature at neutralisation. Suggest a change that could be made to the experiment. **(2 marks)**

(i) The sodium hydroxide solution used was made by dissolving 40.0 g of sodium hydroxide in water and making the solution up to 250 cm³ with water. Calculate the concentration of the solution in mol/dm³.

(2 marks)

(Total for Question 4 is 17 marks)

5 Bath crystals contain Epsom salts, which are hydrated magnesium sulfate crystals. Magnesium sulfate crystals can be prepared in the laboratory by reacting magnesium carbonate and sulfuric acid.

The equation for the reaction is:

$$MgCO_3 + H_2SO_4 \rightarrow MgSO_4 + H_2O + CO_2$$

(a) What is observed in this reaction? **(1 mark)**

(b) Describe how a sample of magnesium sulfate crystals could be made from magnesium carbonate and dilute sulfuric acid. **(4 marks)**

(c) Suggest one safety precaution that should be followed. **(1 mark)**

(d) Calculate the maximum mass of magnesium sulfate that could be made when 2.1 g of magnesium carbonate is reacted with excess sulfuric acid. **(4 marks)**

(e) The student obtained 1.8 g of magnesium sulfate. Calculate the percentage yield. **(2 marks)**

(f) Suggest why the percentage yield is not 100% in this reaction. **(1 mark)**

(Total for Question 5 is 13 marks)

6 A new form of carbon, a fullerene, shown in Figure 5, was discovered in 1985. The first molecules to be isolated had the formula C_{60}, and several other molecules have been discovered since then.

Scientists have been researching and developing uses for fullerenes. One area of research uses these nanoparticles to deliver drugs to different parts of the body. The cage-like molecules contain the drug and are absorbed as they can easily pass through the walls of human cells.

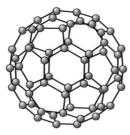

Figure 5

(a) Why can nanoparticles pass easily through cell walls? **(1 mark)**

(b) Suggest one possible problem with the use of nanoparticles in medicines. **(1 mark)**

(c) Name a different form of carbon and describe how its structure is different from fullerene. **(2 marks)**

(Total for Question 6 is 4 marks)

7 A student placed 25.0 cm³ of white wine, containing tartaric acid, in a conical flask. A titration was carried out to find the volume of 0.100 mol/dm³ sodium hydroxide solution needed to neutralise the tartaric acid in the white wine.

(a) Name a suitable indicator for this titration and give the colour change that would be seen. **(2 marks)**

(b) Suggest why this titration is suitable for white wine, but it is not used to find the concentration of acid in red wine. **(1 mark)**

(c) The student carried out four titrations. Her results are shown in the table below.

	Titration 1	Titration 2	Titration 3	Titration 4
Volume of 0.100 mol/dm³ NaOH in cm³	20.05	19.45	18.90	19.00

Concordant results are within 0.10 cm³ of each other.

(i) Use the student's concordant results to work out the mean volume of 0.100 mol/dm³ sodium hydroxide added. **(2 marks)**

The equation for the reaction of tartaric acid in the white wine with the sodium hydroxide is:

$$C_4H_6O_6 + 2NaOH \rightarrow C_4H_4O_6Na_2 + 2H_2O$$

(ii) Calculate the concentration, in mol/dm³, of the tartaric acid. Give your answer to two significant figures. **(5 marks)**

(Total for Question 7 is 10 marks)

8 Methane burns in oxygen. The equation for the reaction is:

$$CH_4 + 2O_2 \rightarrow CO_2 + 2H_2O$$

(a) Complete the reaction profile in Figure 6.

Draw labelled arrows to show

- the energy given out, ΔH
- the activation energy.

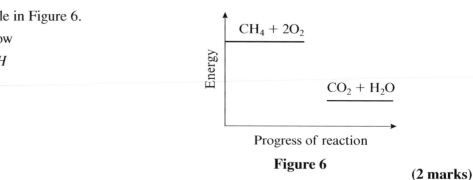

Figure 6

(2 marks)

(b) The overall energy change for this reaction is −698 kJ/mol. Use this information and the information in the table below and Figure 7 to calculate the bond energy for the C–H bond in methane.

Bond	C=O	O=O	O–H
Bond energy in kJ/mol	743	496	463

(3 marks)

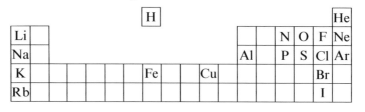

Figure 7

(c) A sample of 0.2 mol carbon dioxide is collected when a sample of methane is burned. Calculate the volume and mass of carbon dioxide gas present in 0.2 moles at room temperature and pressure.

(4 marks)

(Total for Question 8 is 9 marks)

9 This question is about the periodic table and some of its elements.

In 1864, John Newlands arranged all the elements known at the time into a table in order of relative atomic mass. When he did this, he found that each element was similar to the element eight places further on. This repeating pattern of properties was called the law of octaves, but it had problems. For example, iron was in the same group as oxygen and sulfur, which are both non-metals.

(a) Describe how Mendeleev improved on Newland's table and compare Mendeleev's table with the periodic table we use today.

(6 marks)

(b) A part of the periodic table is shown in Figure 8.

																H									He
Li																				N	O	F	Ne
Na														Al		P	S	Cl	Ar
K						Fe		Cu					Br	
Rb													I	

Figure 8

(i) What is the formula of the compound formed between the most reactive element in Group 1 and the most reactive element in Group 7? Use only the elements shown in Figure 8.

☐ LiF   ☐ RbI

☐ LiI   ☐ RbF   **(1 mark)**

(ii) Name a coloured, diatomic gas at room temperature and atmospheric pressure that is present in Figure 8.

☐ bromine

☐ chlorine

☐ helium

☐ nitrogen **(1 mark)**

(c) Potassium reacts with water.

(i) What are the products of this reaction?

☐ potassium hydroxide + hydrogen

☐ potassium hydroxide + oxygen

☐ potassium oxide + hydrogen

☐ potassium oxide + oxygen **(1 mark)**

(ii) Write a half-equation for the formation of a potassium ion from potassium. **(2 marks)**

(d) Potassium reacts with chlorine to form potassium chloride.

(i) What are the properties of potassium chloride?

☐ coloured solid, soluble in water

☐ coloured solid, insoluble in water

☐ white solid, soluble in water

☐ white solid, insoluble in water **(1 mark)**

(ii) What are the products of the electrolysis of molten potassium chloride?

Product at cathode	Product at anode
☐ potassium	chlorine
☐ potassium	oxygen
☐ hydrogen	chlorine
☐ hydrogen	oxygen

(1 mark)

(e) Aluminium metal reacts with iron(III) oxide. The equation for the reaction is:

$Fe_2O_3 + 2Al \rightarrow 2Fe + Al_2O_3$

Write an ionic equation for the reaction of aluminium with iron(III) oxide. **(2 marks)**

(Total for Question 9 is 15 marks)

Practice Paper 2

Time allowed: 1 hour 45 minutes

Total marks: 100

AQA publishes official Sample Assessment Material on its website. This practice exam paper has been written to help you practise what you have learned and may not be representative of a real exam paper.

1 Some substances are pure.

(a) Which statement is always true for a pure substance?

☐ It always boils at 100 °C.

☐ It contains only one type of atom.

☐ It has a sharp melting point.

☐ It is solid at room temperature. **(1 mark)**

(b) Potable water can be produced by desalination of salty water. Which is a method of desalination?

☐ bubbling ozone into the water

☐ reverse osmosis

☐ filtration

☐ sedimentation **(1 mark)**

(c) Potable water can also be produced from groundwater. Which is the method used?

☐ anaerobic digestion of sewage sludge in the water, followed by sterilisation

☐ anaerobic biological treatment of effluent, followed by sterilisation

☐ sterilising the water with chlorine followed by passing through filter beds

☐ passing the water through filter beds and sterilising with ozone **(1 mark)**

(Total for Question 1 is 3 marks)

2 (a) The table below shows some tests that were carried out on a solution of calcium iodide.
Complete the table.

Test	Observation
(i) Flame test	
(ii) 6 drops of sodium hydroxide solution were added, followed by excess sodium hydroxide solution.	
(iii) Some dilute nitric acid and silver nitrate solution were added.	

(4 marks)

(b) Another solution, containing a different metal ion, gave the same results for tests (ii) and (iii).
Suggest the formula of the substance dissolved in this solution. **(1 mark)**

(c) Give two advantages of flame emission spectroscopy to identify metal ion solutions. **(2 marks)**

(d) Some dilute hydrochloric acid was added to a sample of sodium carbonate. Describe how you would test for the gas produced. **(2 marks)**

(Total for Question 2 is 9 marks)

3 Long-chain alkanes can be cracked to form short-chain alkenes. The apparatus in Figure 1 was used to produce ethene.

(a) Which essential piece of apparatus is missing from the diagram? **(1 mark)**

(b) Suggest why the first tube of gas collected should be discarded. **(1 mark)**

(c) What is the function of the aluminium oxide? **(1 mark)**

(d) Describe a chemical test to distinguish an alkane from an alkene. **(2 marks)**

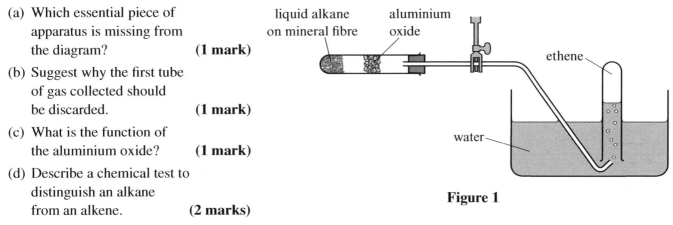

Figure 1

(e) Complete the following table.

Name of homologous series	General formula	Molecular formula of compound with three carbon atoms
alkanes		C_3H_8
alkenes	C_nH_{2n}	

(2 marks)

(f) Compare the reactions of ethene with hydrogen, water and chlorine, giving the type of each reaction, the conditions, and the names and displayed structural product of each reaction. **(6 marks)**

(g) Figure 2 shows the structure of four organic compounds, A, B, C and D.

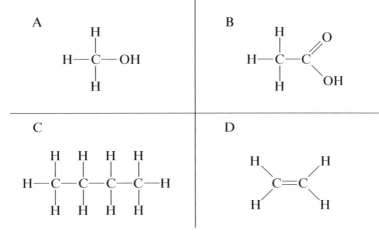

Figure 2

(i) Name compound A. **(1 mark)**

(ii) Why is C a hydrocarbon? **(1 mark)**

(iii) Which compound, A, B, C or D, reacts with sodium carbonate? **(1 mark)**

(iv) Which compound, A, B, C or D, is unsaturated? **(1 mark)**

(v) B is formed by the oxidation of which organic compound?

☐ ethane

☐ ethene

☐ ethanol

☐ polyester **(1 mark)**

(vi) Name the addition polymer that can be formed from D. **(1 mark)**

(Total for Question 3 is 19 marks)

4 The Earth's early atmosphere consisted mainly of carbon dioxide with little or no oxygen gas. Today the atmosphere contains about 21% oxygen gas.

(a) Name the gas that makes up most of the atmosphere today. **(1 mark)**

(b) Explain how the amount of oxygen in the atmosphere increased. Include a balanced equation in your explanation. **(4 marks)**

(c) Copper powder was heated strongly in a test tube. Figure 3 shows a diagram of the apparatus used. The copper reacted with oxygen in the air in the apparatus. Heating was stopped when there was no further change in the reading on the gas syringe.

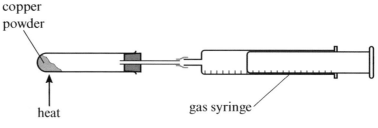

copper powder

heat

gas syringe

Figure 3

(i) Write a balanced equation for the reaction that occurred in the test tube. **(2 marks)**

(ii) Why should the gas be left for a few minutes before reading the volume left at the end?

☐ Reading the volume while the apparatus is hot is dangerous.

☐ The apparatus must be left to allow the reaction to finish.

☐ The gas must be at room temperature when its volume is measured.

☐ The copper expands when it is hot. **(1 mark)**

(iii) At the end of the experiment, not all of the copper had reacted. Suggest a reason for this. **(1 mark)**

(iv) The table below shows the results of the experiment.

Initial volume of gas in syringe in cm³	32
Final volume of gas in syringe in cm³	24

Calculate the percentage decrease in the volume of gas from that originally in the syringe. **(3 marks)**

(v) Compare your answer with the percentage of oxygen in today's atmosphere and suggest why it is different. **(2 marks)**

(d) Which pollutant gas is produced by the decomposition of vegetation?

☐ carbon monoxide

☐ methane

☐ nitrogen oxide

☐ nitrogen dioxide **(1 mark)**

(e) Sulfur dioxide is a pollutant gas found in the atmosphere. Describe how sulfur dioxide enters the atmosphere and the problems it causes. **(4 marks)**

(Total for Question 4 is 19 marks)

5 A student investigated the reaction of 0.1 g of magnesium ribbon with 50 cm³ of dilute hydrochloric acid of concentration 1 mol/dm³ at 20 °C. Figure 4 shows the apparatus used.

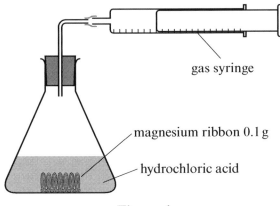

gas syringe

magnesium ribbon 0.1 g

hydrochloric acid

Figure 4

(a) Complete and balance the equation for the reaction between magnesium and hydrochloric acid.

.... + ... → ...H₂ +................................ **(2 marks)**

(b) Give one advantage and one disadvantage of using a measuring cylinder to add the acid to the flask.

(2 marks)

(c) The table below shows the results of this experiment.

Time in s	0	30	60	90	120	150	180
Volume of gas in cm³	0	13	22	30	36	43	49

On Figure 5:

* plot these results on the grid
* draw a line of best fit.

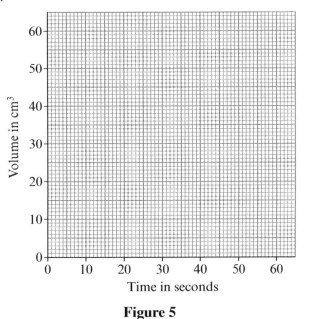

Figure 5

(4 marks)

(d) Use your graph to find the time needed to collect 25 cm³ of gas. **(1 mark)**

(e) Calculate the mean rate of the reaction for the first 30 seconds of the reaction. Give your answer to one significant figure. State the units. **(4 marks)**

(f) Use your graph to determine the rate of reaction at 60 seconds. Show your working on Figure 5. Give your answer in standard form. **(4 marks)**

(g) Suggest one improvement to this experiment. Give a reason for your answer. **(2 marks)**

(h) Sketch a line on the grid in Figure 5 to show the results you would expect if the experiment were repeated using 0.1 g of magnesium filings in 50 cm³ of 1 mol/dm³ hydrochloric acid at 20 °C. Label this line A. **(2 marks)**

(i) Explain how and why the rate of reaction changes if the experiment is repeated at 40 °C. **(3 marks)**

(Total for Question 5 is 24 marks)

6 Solder is an alloy of tin and lead.

(a) A sample of a solder was made by mixing 22.5 g of lead with 15.0 g of tin. Calculate the percentage of tin by mass in this solder. **(3 marks)**

(b) Why are alloys stronger than pure metals?

☐ There are stronger bonds between the molecules they contain.

☐ They combine the properties of the metals from which they are made.

☐ They have atoms of different sizes in their structures.

☐ They are made using electrolysis. **(1 mark)**

(Total for Question 6 is 4 marks)

7 Figure 6 shows the structure of an amino acid, X.

Figure 6

(a) Circle the two functional groups present. **(2 marks)**

(b) X can polymerise, as shown in the equation below:

n H₂N-CH(CH₃)-COOH → -(NH-CH(CH₃)-COO)ₙ- + nH₂O

Name the type of polymerisation shown. **(1 mark)**

(c) Figure 7 shows a chromatogram obtained when a mixture of amino acids was analysed.

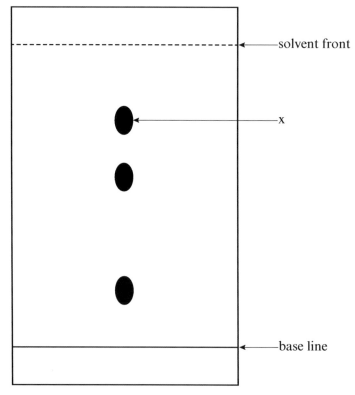

Figure 7

(i) Why must the base line be drawn in pencil instead of pen? **(1 mark)**

(ii) How many amino acids were in the mixture? **(1 mark)**

(iii) Calculate the R_f value for the amino acid X. Use the table below to identify X.

Amino acid	R_f value
alanine	0.38
aspartic acid	0.15
glycine	0.26
leucine	0.75
methionine	0.58

(4 marks)

(Total for Question 7 is 9 marks)

8 A catalytic converter is a device in a car exhaust system, in which a catalyst is spread thinly over a honeycomb structure.

(a) What is a catalyst? **(2 marks)**

(b) The following reversible reaction takes place in a catalytic converter.

$$2CO + 2NO \rightleftharpoons 2CO_2 + N_2$$

The forward reaction is exothermic.

(i) Use Le Chatelier's principle to predict the effect of increasing temperature on the amount of carbon dioxide produced at equilibrium. Give a reason for your prediction. **(2 marks)**

(ii) Explain how increasing the pressure of the reactants will affect the amount of carbon dioxide produced at equilibrium. **(2 marks)**

(Total for Question 8 is 6 marks)

9 New ways of extracting copper include phytomining and bioleaching.

(a) Describe the methods of phytomining and bioleaching, and explain how copper metal is produced using these methods and scrap iron. **(5 marks)**

(b) Give two advantages of phytomining over traditional extraction methods. **(2 marks)**

(Total for Question 9 is 7 marks)

The periodic table of the elements

Key

| relative atomic mass |
| **atomic symbol** |
| name |
| atomic (proton) number |

1	2												3	4	5	6	7	0
								1 **H** hydrogen 1										4 **He** helium 2
7 **Li** lithium 3	9 **Be** beryllium 4												11 **B** boron 5	12 **C** carbon 6	14 **N** nitrogen 7	16 **O** oxygen 8	19 **F** fluorine 9	20 **Ne** neon 10
23 **Na** sodium 11	24 **Mg** magnesium 12												27 **Al** aluminium 13	28 **Si** silicon 14	31 **P** phosphorus 15	32 **S** sulfur 16	35.5 **Cl** chlorine 17	40 **Ar** argon 18
39 **K** potassium 19	40 **Ca** calcium 20	45 **Sc** scandium 21	48 **Ti** titanium 22	51 **V** vanadium 23	52 **Cr** chromium 24	55 **Mn** manganese 25	56 **Fe** iron 26	59 **Co** cobalt 27	59 **Ni** nickel 28	63.5 **Cu** copper 29	65 **Zn** zinc 30		70 **Ga** gallium 31	73 **Ge** germanium 32	75 **As** arsenic 33	79 **Se** selenium 34	80 **Br** bromine 35	84 **Kr** krypton 36
85 **Rb** rubidium 37	88 **Sr** strontium 38	89 **Y** yttrium 39	91 **Zr** zirconium 40	93 **Nb** niobium 41	96 **Mo** molybdenum 42	[98] **Tc** technetium 43	101 **Ru** ruthenium 44	103 **Rh** rhodium 45	106 **Pd** palladium 46	108 **Ag** silver 47	112 **Cd** cadmium 48		115 **In** indium 49	119 **Sn** tin 50	122 **Sb** antimony 51	128 **Te** tellurium 52	127 **I** iodine 53	131 **Xe** xenon 54
133 **Cs** caesium 55	137 **Ba** barium 56	139 **La*** lanthanum 57	178 **Hf** hafnium 72	181 **Ta** tantalum 73	184 **W** tungsten 74	186 **Re** rhenium 75	190 **Os** osmium 76	192 **Ir** iridium 77	195 **Pt** platinum 78	197 **Au** gold 79	201 **Hg** mercury 80		204 **Tl** thallium 81	207 **Pb** lead 82	209 **Bi** bismuth 83	[209] **Po** polonium 84	[210] **At** astatine 85	[222] **Rn** radon 86
[223] **Fr** francium 87	[226] **Ra** radium 88	[227] **Ac*** actinium 89	[261] **Rf** rutherfordium 104	[262] **Db** dubnium 105	[266] **Sg** seaborgium 106	[264] **Bh** bohrium 107	[277] **Hs** hassium 108	[268] **Mt** meitnerium 109	[271] **Ds** damstadtium 110	[272] **Rg** roentgenium 111	[285] **Cn** copernicium 112		[286] **Uut** ununtrium 113	[289] **Fl** flerovium 114	[289] **Uup** ununpentium 115	[293] **Lv** livermorium 116	[294] **Uus** ununseptium 117	[294] **Uuo** ununoctium 118

* The Lanthanides (atomic numbers 58 – 71) and the Actinides (atomic numbers 90 – 103) have been omitted.

Relative atomic masses for **Cu** and **Cl** have not been rounded to the nearest whole number.

Answers

Extended response questions

Answers to 6-mark questions are indicated with a star (*).

In your exam, your answers to 6-mark questions will be marked on how well you present and organise your response, not just on the scientific content. Your responses should contain most or all of the points given in the answers below, but you should also make sure that you show how the points link to each other, and structure your response in a clear and logical way.

1. Elements, mixtures and compounds

1 a substance made up of two or more elements chemically joined together (1)

2 (a) Na (1)   (b) NH_3 (1)   (c) NaOH (1)

3 (a) iron sulfide (1)

  (b) In the mixture, the iron powder will cluster on the magnet (1), while the sulfur will remain behind (1). However, compounds can only be separated by chemical reactions, so the magnet will affect all of the powdered compound in the same way. (1)

4 (a) A (1)   (b) C (1)   (c) B (1)

2. Filtration, crystallisation and chromatography

1 (a) A crystallisation (1), B chromatography (1), C filtration (1)

  (b) (i) C, (ii) A, (iii), C (iv) B (4)

2 *Step 1* addition of water

  Reason: to dissolve the sodium chloride (1)

  Step 2 heating and stirring

  Reason: to speed up dissolving/ensure the salt fully dissolves (1)

  Step 3 filtration

  Reason: to separate the insoluble sand from the salt solution (1)

  Step 4 evaporation

  Reason: to remove most of the water (1)

    (1 mark for correct order of steps)

3. Distillation

1 (a) distillation (1)

  (b) to condense the vapour (1)

  (c) evaporation (1)

  (d) C fractionating column (1), D water in (1), E thermometer (1)

  (e) Use an electrical heating mantle. (1)

2 The salty water is heated and the water boils and evaporates (1). Water vapour moves up and into the condenser, while the salt solution in the flask grows more concentrated (1). The condenser cools the water vapour and it condenses to pure water in the beaker. (1)

4. Historical models of the atom

1 (a) positive (1)

  (b) electron (1)

2 From the fact that most of the positive particles passed straight through the foil without deflection, they deduced that most of the atom is made up of space (1), but that there must be small concentrated charges in the atom (1). The nucleus, like the alpha particle, is positively charged. (1)

3 (a) (James) Chadwick (1)

  (b) It is electrically neutral (1), so is not deflected by electric or magnetic fields. (1)

4 The plum pudding model suggested that an atom was a ball of positive charge. (1)

  The nuclear model has positive protons in a nucleus. (1)

  In the plum pudding model, the negative electrons are embedded throughout the sphere of positive charge (1). In the nuclear model, electrons orbit the nucleus at specific distances. (1)

5. Particles in an atom

1 (a) The total number of the protons and neutrons in an atom is its mass number. (1)

  (b) The sodium atom has an equal number of protons and electrons so the total number of positive charges cancels the total number of negative charges. (1)

  (c) 11 protons, 12 neutrons, 11 electrons (3)

  (d) protons and neutrons (1) **(both needed for mark)**

2 (a)

Atom	Atomic number	Mass number	Number of electrons	Number of neutrons	Number of protons
A	27	59	27	59 − 27 = 32	27
B	28	59	28	31	28
C	13	27	13	14	13
D	19	39	19	20	19

    (4) (1 for each correct line)

  (b) A cobalt, B nickel, C aluminium, D potassium (4)

3 The lithium atom is made up of a central nucleus containing 3 protons (1) and 4 neutrons (1) **(either order)**. Around the nucleus there are 3 electrons. (1)

6. Atomic structure and isotopes

1 (a)

Particle	Relative mass	Relative charge
electron	very small	−1
neutron	1	0
proton	1	+1

    (3) (1 for each correct line)

  (b) 19 protons (1), 20 neutrons (1), 19 electrons (1)

  (c) 1×10^{-10} m (1), 0.1 nm (1)

  (d) They have the same number of protons (atomic number) (1), but different numbers of neutrons (mass number). (1)

2 (a) They have the same electronic structure. (1)

  (b) ^{12}C and ^{13}C atoms both have 6 electrons and 6 protons (1). ^{12}C atoms have 6 neutrons and ^{13}C atoms have 7 neutrons. (1)

  (c) $= \dfrac{12 \times 99 + 13 \times 1}{99+1} = 12.01 = 12.0$ (to one decimal place) (2)

7. Electronic structure

1 The number of electrons in each shell (from the middle outwards) should be: 2,8,3 (1); 2,8,7 (1); 2,8,8,2. (1)

2 (a) silicon (1), atomic number 14 (1)

  (b) 14 protons and 14 electrons (1)

  (c) 14 (mass number 28 − atomic number 14) (1)

3 (a) 2,8,8,1 (1)   (b) 2,8,5 (1)

4 Missing name: neon (1)

  The missing atomic numbers are 3 and 10. (1)

  The electron arrangements in the blank diagrams are 2,8,3 and 2,8 (1).

  The missing electronic structures are 2,1 and 2,8,3. (1)

8. Development of the periodic table

1 atomic weight (1)

2 (a) The early tables were incomplete as some elements were unknown (1) and some elements were placed in inappropriate groups because the order of atomic weights does not perfectly match the order of atomic numbers. (1)

  (b) He left gaps for elements that he thought had not yet been discovered (1), and in some places he changed the order despite the atomic weights in order to make chemical properties match in a group. (1)

  (c) It has similar chemical properties. (1)

  (d) germanium (1)

3 Mendeleev listed elements in his periodic table by atomic weight but the modern periodic table lists elements by atomic number (1). In Mendeleev's periodic table there is no group 0 and no block of transition metals, but in the modern periodic table both are present (1). In Mendeleev's periodic table there are gaps for undiscovered elements, but in the modern periodic table there are no spaces (1). Mendeleev's periodic table has fewer elements than the modern periodic table. (1)

9. The modern periodic table

1 by increasing atomic number (1)

2 (a) A and F (1) (b) D (1) (c) E (1)

3 (a) They all have the same number of electrons in the outer shell of their atoms. (1)

(b) The others are metals. (1)

(c) The element with the smallest atomic number is boron. As the atomic number gives the number of protons, boron has the lowest number of protons. (1)

10. Group 0

1 2 (1)

2 (a) $-160\,°C$ (accept $-145\,°C$ to $-165\,°C$) (1)

(b) The boiling point increases with increasing relative atomic mass. (1)

(c) $1.7\,g/dm^3$ (accept 1.6–$1.8\,g/dm^3$) (1)

(d) 2; 2,8 (2)

(e) They both have a stable arrangement of electrons (a full outer electron shell). (1)

3 Their atoms have stable arrangements of electrons. (1)

4 (a) 18 (1)

(b) argon (1)

(c) 22 (1)

(d) It has a stable arrangement of electrons. (1)

11. Group 1

1 (a) rubidium (1)

(b) rubidium (1)

2 (a) The compounds formed between alkali metals and non-metals are ionic, soluble, and white. (3)

(b) The reactivity increases down the group (1) because the outer electron is further from the nucleus (1) and there is less attraction between the outer electron and the nucleus, therefore the outer electron is lost more easily (shielding). (1)

3 (a) Their atoms have one electron in the outer shell. (1)

(b) Similarities (any two from): metal floats, moves, disappears, bubbles (1)

Differences (any two from): potassium reacts faster/bubbles faster, moves faster (on the surface), melts, produces (lilac/purple) flame (1)

Products: lithium hydroxide and hydrogen gas (1), potassium hydroxide and hydrogen gas (1)

4 $2Na + 2H_2O \rightarrow 2NaOH + H_2$ (1)

Ion: hydroxide ion (1)

12. Group 7

1 As you go down the group the melting point increases. (1)

This is because the forces between larger molecules are stronger. (1)

2 (a) (i) Halogens are toxic. (1)

(ii) Fluorine is so reactive that the reaction with iron is dangerous. (1)

(b) iron(III) chloride (1), iron(II) bromide (1)

(c) $Fe + I_2 \rightarrow FeI_2$ (2)

(d) The atoms are larger and the outer shell is further from the nucleus (1). The nuclear charge is less attractive to more distant electrons and so they are gained less easily in reactions. (1)

3 (a) $Cl_2(g) + 2NaBr(aq) \rightarrow Br_2(aq) + 2NaCl(aq)$ (1 mark for formulae, 1 mark for balanced numbers)

(b) displacement or redox (1)

(c) $2Br^- + Cl_2 \rightarrow Br_2 + 2Cl^-$ (1)

13. Transition metals

1 false (1), true (1), false (1), false (1)

2 (a) B (1) (b) A (1) (c) B (1) (d) E (1)

3 (a) The metallic ion in copper(II) sulfate is copper, which has a 2+ charge. Sulfate has a charge of 2−, hence the formula of copper(II) sulfate is $CuSO_4$ (1). In sodium hydrogen carbonate, the metallic ion is sodium, which has a 1+ charge, and hydrogen carbonate has a charge of 1−. Hence, the formula is $NaHCO_3$. (1)

(b) Copper(II) sulfate is coloured and sodium hydrogen carbonate is white. (2)

(c) Two from: strong/hard, not very reactive, high melting point (2)

14. Chemical equations

1 (a) One molecule of methane reacts with two molecules of oxygen to produce one molecule of carbon dioxide and two molecules of water. (2)

(b) 16 g (1)

2 (a) $C_2H_6 + 3.5O_2 \rightarrow 2CO_2 + 3H_2O$ (equivalent to $2C_2H_6 + 7O_2 \rightarrow 4CO_2 + 6H_2O$) (2)

(b) $C_3H_8 + 5O_2 \rightarrow 3CO_2 + 4H_2O$ (2)

(c) carbon dioxide and water (2)

3 (a) $4Na + O_2 \rightarrow 2Na_2O$ (2)

(b) $4K + O_2 \rightarrow 2K_2O$ (2)

4 (a) $2HCl + Ca \rightarrow CaCl_2 + H_2$ (1)

(b) $MgCO_3 + 2HCl \rightarrow MgCl_2 + H_2O + CO_2$ (1)

(c) $2Al + 6HCl \rightarrow 2AlCl_3 + 3H_2$ (1)

5 (a) $2Na + Cl_2 \rightarrow 2NaCl$ (2)

(b) $2K + 2H_2O \rightarrow 2KOH + H_2$ (2)

(c) $Mg + 2HCl \rightarrow MgCl_2 + H_2$ (2)

15. Extended response – Atomic structure

Answer could include the following points.

Physical properties:

- Both transition metals and alkali metals are metals that have shiny surfaces when fresh and they conduct electricity.
- Transition metals have high melting points.
- Alkali metals have low melting points.
- Transition metals have high densities.
- Alkali metals have low densities/are less dense than water.
- Transition metals are hard (used in building and engineering applications, e.g. iron, chromium).
- Alkali metals are soft (can be cut with a knife).

Chemical properties:

- Transition metals have low reactivity/react slowly with water (e.g. iron rusting).
- Alkali metals are very reactive/react quickly with water (e.g. potassium forming potassium hydroxide).
- Both form positive ions.
- Transition metal ions can form several different ions with different charges (e.g. iron(II) oxide and iron(III) oxide or copper(I) oxide and copper(II) oxide).
- Alkali metals form ions with a charge of +1 because they have only one outer electron (e.g. K^+ in KOH).
- Transition metal compounds are often coloured.
- Alkali metal compounds are white.

This answer is not exhaustive; other creditworthy responses will be awarded marks too.

16. Forming bonds

1 carbon and oxygen (1)

2 (a) ionic (1)

(b) metallic (1)

(c) covalent (1)

(d) ionic (1)

3 (a) metallic (1)

(b) delocalised electrons (1)

(c) metal ions/cations (1)

4 Ionic bonding is the attraction between oppositely charged ions (1). Covalent bonding is the sharing of electrons between atoms. (1)

5 (a) NH_3 (1)

(b) covalent bonding, sharing of electrons between atoms (1)

17. Ionic bonding

1 Each chlorine atom gains one electron. (1)

2

Atom/ion	Number of protons	Electronic configuration
Al	13	2,8,3
S^{2-}	16	2,8,8
Na^+	11	2,8
Mg^{2+}	12	2,8

(3) (1 for each correct column)

3 Two sodium atoms each lose one electron (1) forming Na^+ ions (1). These electrons are transferred to the oxygen atom (1), so the oxygen atom gains 2 electrons (1) forming O^{2-} ions. (1)

4 (a) oxide ion (1)

(b) potassium ion (1)

(c) aluminium ion (1)

(d) potassium ion (1)

18. Giant ionic lattices

1 six chloride ions (1)

2 (a) ionic (1)

(b) giant ionic lattice (1)

(c) It shows gaps between the ions, but in the crystal the ions are touching. (1)

3 (a) Any two from:

- It shows the particles as the same size, but in sodium chloride the ions are different sizes.
- The charges are not shown.
- The model is only two-dimensional. (2)

(b)

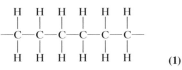

**(1 mark for outer shell of atoms,
1 mark for outer shell of ions, 1 mark
for each charge = 4 marks)**

(c) The ions are held together strongly by
the attraction of opposite ions **(1)**. Much
energy is needed to break this strong
bond. **(1)**

19. Covalent bonding

1 (a) The lone pair is the ×× and the covalent
bond any ×• **(2)**

(b) PH_3 **(1)**

(c) compound **(1)**

2

Oxygen	Water	Nitrogen
O_2	H_2O	N_2

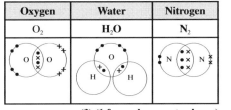

(3) (1 for each correct column)

3 (a) a shared pair of electrons **(1)**

(b)

H, H, Si, H, H

(2)

20. Small molecules

1 (a) (i) CCl_4 **(1)**

(ii) HCl **(1)**

(b)

Cl, C, Cl, Cl, Cl

(1)

(c)

H, Cl

(1)

(d) The forces between the molecules are
stronger. **(1)**

2 B **(1)**

21. Polymer molecules

1 large; atoms; strong; strong **(4)**

2 (a) a large number of repeating units **(1)**

(b) covalent bond **(1)**

(c)

H H H H H H
—C—C—C—C—C—C—
H H H H H H

(1)

(d) The intermolecular forces between the
polymer molecules are stronger **(1)**, as
polythene is a much larger molecule than
ethene. **(1)**

3

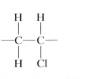

H H
—C—C—
H Cl

(1)

22. Diamond and graphite

1 (a) simple molecular **(1)**

(b) simple molecular **(1)**

(c) giant covalent **(1)**

(d) simple molecular **(1)**

2 (a) The atoms are linked to others by strong
covalent bonds **(1)** in a giant structure **(1)**.
A lot of energy is needed to break these
strong bonds. **(1)**

(b) Each carbon atom forms four strong
covalent bonds **(1)** with other carbon
atoms **(1)** in a giant covalent structure
(macromolecules). **(1)**

3 covalent bonds **(1)**

4 A carbon atom, B covalent bond, C
intermolecular force **(3)**

5 It contains delocalised electrons that are free
to move. **(1)**

23. Graphene and fullerenes

1 graphene **(1)**

2 (a) 5, 6, 7 **(2)**

(b) carbon nanotubes and
buckminsterfullerene **(2)**

(c) They are hollow **(1)**, so a drug can
be carried inside them **(1)** (and so get
through cell walls).

3 Graphene is a single layer of graphite that is
one atom thick. **(2)**

4 (a) carbon nanotube **(1)**

(b) good **(1)**; high **(1)**

(c) Nanotubes can slide smoothly over each
other **(1)**, because there are no covalent
bonds between the nanotubes. **(1)**

24. Metallic bonding

1 (a) delocalised electron **(1)**

(b) It has layers of positive ions **(1)**
surrounded by delocalised electrons. **(1)**

(c) It is the electrostatic attraction **(1)** between
positive metal ions and delocalised
electrons. **(1)**

2 (a) (i) metallic **(1)**

(ii) ionic **(1)**

(b)

*Answer could include the following points.

• Calcium conducts as solid and molten;
calcium chloride only conducts when
molten.

  o The conductivity of solid calcium is
due to the presence of delocalised
electrons that can move and carry
charge.

  o Solid calcium chloride does not
conduct as the ions cannot move, but
when molten the ions can move and
carry charge.

• Calcium and calcium chloride both have
high melting points.

  o In calcium, this is because the strong
metallic bonds need a lot of energy
to break.

  o In calcium chloride, this is because
the strong ionic bonds need a lot of
energy to break.

This answer is not exhaustive; other creditworthy
responses will be awarded marks too.

25. Giant metallic structures and alloys

1 (a) (i) The layers of atoms can slide over
each other **(1)** but the attraction
between electrons and positive ions
holds the structure together. **(1)**

(ii) protons 79 **(1)**, neutrons 118 **(1)**

(b) In an alloy there are atoms of different
sizes present **(1)**. This makes it more
difficult for the layers of atoms to slide
over each other than in a pure metal. **(1)**

(c) $2Au + 3Cl_2 \rightarrow 2AuCl_3$ **(2)**

2 Metals contain delocalised electrons, which
can move and transfer energy. **(1)**

26. The three states of matter

1 (a) The particles are regularly arranged, close
together **(1)** and vibrate about a fixed
position. **(1)**

(b) The particles gain energy and vibrate
faster. They overcome the forces between
them and move slightly apart so that they
are able to move more freely, away from
their fixed positions **(1)**; this state is a
liquid. **(1)**

(c) Any two from:

• There are no forces between the spheres,
unlike real particles.

• All particles are represented as spheres,
regardless of their shape.

• The spheres are solid and inelastic, which
is not the case for particles. **(2)**

2 (a) (i) B **(1)**

(ii) C **(1)**

(iii) D **(1)**

(b) C is sodium chloride, which has strong
ionic bonds between the ions **(1)**, so a lot
of energy is needed to break these and
make the solid melt **(1)**. A is a liquid at
room temperature and does not conduct
electricity, so it has a simple molecular
structure. Therefore, little energy is
needed **(1)** to overcome the weak forces
between the molecules. **(1)**

27. Nanoscience

1 (a) a particle with a diameter 1–100 nm **(1)**

(b) Compared with the particles in ordinary
powders, nanoparticles have a surface
area that is very large. **(1)**

2 (a) Any one from:

• Nanoparticles may have properties
different from those for the same
materials in bulk because of their high
surface-area-to-volume ratio.

• Smaller quantities are needed to be
effective than for materials with normal
particle sizes. **(1)**

(b) large surface area **(1)**

(c) We still do not know the risks involved. **(1)**

3 (a) atom 2×10^{-10} m; nanoparticle $1–100 \times 10^{-9}$ m **(2)**

(b) The surface-area-to-volume ratio is larger for the smaller particle by a factor of ten. **(1)**

28. Extended response – Bonding and structure

*Answer could include the following points.

- Metals have giant structures in which positive ions are surrounded by delocalised electrons.
- The electrical attraction between the ions and electrons holds them together.
- This metallic bonding is strong and the bonds take a lot of energy to break, so they have high melting points.
- Metals are good conductors of electricity because the delocalised electrons in the metal move and carry electrical charge.
- Metals are good conductors of heat because thermal energy is carried by the delocalised electrons.
- The layers of metal particles are able to slide over each other, but they are held in the structure by the attraction between the electrons and ions. This makes copper soft and malleable.

This answer is not exhaustive; other creditworthy responses will be awarded marks too.

29. Relative formula mass

1 (a) 62 **(1)**

(b) 342 **(1)**

(c) 88 **(1)**

(d) 98 **(1)**

(e) NO_3^- = 62, so $Ca(NO_3)_2$ = 40 + 62 × 2 = 164 **(1)**

(f) SO_4^{2-} = 96, so $Al_2(SO_4)_3$ = 342 **(1)**

2 (a) 34 **(1)**

(b) 2 × 98 = 196 **(1)**

(c) 0.5 × 58 = 29 **(1)**

30. Moles

1 6.02×10^{24} **(1)**

2 (a) 88/44 = 2 mol **(2)**

(b) 4/32 = 0.125 mol **(2)**

(c) 7.4/74 = 0.1 mol **(2)**

(d) 17.1/342 = 0.05 mol **(2)**

3 (a) M_r of $Ca(CO_3)$ = 100 **(1)**; mass = 0.25 × 100 = 25 g **(1)**

(b) M_r of $Mg(NO_3)_2$ = 148 **(1)**; mass = 1.2 × 148 = 177.6 g **(1)**

31. Balanced equations, moles and masses

1 (a) (i) sodium hydrogen carbonate + citric acid → sodium citrate + carbon dioxide + water **(1)**

(ii) carbon dioxide gas **(1)**

(b) (i) flask on balance **(1)**, conical flask with cotton wool stopper both labelled **(1)**; indigestion tablet + 50 cm³ water in flask, both labelled **(1)**

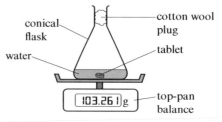

conical flask
cotton wool plug
water
tablet
top-pan balance
103.261 g
(3)

(ii) Carbon dioxide gas is produced **(1)** and escapes the conical flask through the cotton wool plug. **(1)**

(iii) to prevent droplets of spray being lost and affecting the balance reading **(1)**

(iv) The balance reading stays the same **(1)** as no gas is produced and so nothing leaves or enters the flask. **(1)**

2 Total formula mass of reactants = 24 + (2 × [35.5 + 1]) = 97 **(1)**

Total formula mass of products = (24 + [2 × 35.5]) + (2 × 1) = 97 **(1)**

3 (a) 30.24 − 30.00 = 0.24 g **(1)**

(b) 30.40 − 30.00 = 0.40 g **(1)**

(c) 0.40 − 0.24 = 0.16 g **(1)**

32. Reacting masses

1 moles of Mg = 1.2/24 = 0.05, so moles of Cu = 0.05

mass of Cu = 0.05 × 63.5 = 3.2 g **(1)**

2 M_r $Cu(NO_3)_2$ = 63.5 + 2 × (14 + [3 × 16]) = 187.5 **(1)**

Therefore, moles $Cu(NO_3)_2$ = 9.4/187.5 = 0.05 **(1)**

Ratio $1CuCO_3$: $1Cu(NO_3)_2$. So, 0.05 moles of $CuCO_3$ are required. **(1)**

Mass of $CuCO_3$ = 0.05 × 123.5 = 6.175 = 6.2 g **(1)**

3 Moles of $KMnO_4$ used = 5.53/158 = 0.035 **(1)**

Moles of O_2 produced = 0.035/2 = 0.0175 **(1)**

Mass of O_2 = 0.0175 × 32 **(1)** = 0.56 g **(1)**

4 10 kg = 10 000 g **(1)**

Moles of iron(III) oxide = 10 000/160 = 62.50 **(1)**

Ratio $1Fe_2O_3$: 2Fe

Moles of Fe = 125.0 **(1)**

Mass of Fe = moles × M_r = 125.0 × 56 = 7000 g **(1)**

33. Further reacting masses

1 (a) (i) Moles of chlorine = 10.65/71 = 0.15 **(2)**

(ii) Moles of ethanoic acid = 6.0/60 = 0.10 **(2)**

(b) Ratio is $3Cl_2$: $1CH_3COOH$, so 0.15 moles Cl_2 would react with 0.05 moles CH_3COOH. **(1)**

The ethanoic acid is in excess and the limiting reactant is chlorine. **(1)**

(c) Ratio is $3Cl_2$: $1CCl_3COOH$, so 0.15 moles Cl_2 produces 0.05 moles CCl_3COOH. **(1)**

0.05 moles CCl_3COOH = 0.05 × 163.5 = 8.2 g **(1)**

2 Moles NH_4Cl = 20/53.5 = 0.374 **(1)**

Moles CaO = 50/56 = 0.893 **(1)**

0.374/2 < 0.893; calcium oxide is in excess and ammonium chloride is the limiting reactant.

2 mol NH_4Cl : 1 mol $CaCl_2$; 0.374 mol : 0.187 mol **(1)**

Mass = 0.187 × 111 = 20.757 **(1)** = 20.8 g to three significant figures **(1)**

34. Concentration of a solution

1 number of moles = 4/40 = 0.1 mol in 1 dm³ = 0.1 mol/dm³ **(1)**

2 (a) 0.2 × 4 = 0.8 mol/dm³ **(1)**

(b) 3/2 = 1.5 mol/dm³ **(1)**

(c) 2.1/84 = 0.025 moles in 250 cm³

0.025 × 4 = 0.1 mol/dm³ **(2)**

(d) 4.9/98 = 0.05 moles in 500 cm³ = 0.1 mol/dm³ **(2)**

3 (a) Number of moles = 25 × 0.5/1000 = 0.0125 **(2)**

(b) 10 × 0.2/1000 = 0.002 **(2)**

(c) 0.1/4 = 0.025 moles in 1 dm³

0.025 × 98 = 2.45 g **(2)**

35. Core practical – Titration

1 (a) A burette **(1)**, B conical flask **(1)**, C white tile **(1)**

(b) volumetric pipette **(1)**

(c) Add 3 drops of indicator to the potassium hydroxide solution, for example, phenolphthalein/methyl orange/litmus **(1)**. Add acid (from the burette) **(1)** in a dropwise manner towards the end point, until the indicator just changes colour **(1)**, from pink to colourless (for phenolphthalein), yellow to red (for methyl orange) or blue to red (for litmus). **(1)**

(d) to help see the indicator colour change clearly **(1)**

(e) (i) 27.20; 26.75; 26.05; 26.15 **(4)**

(ii) $\dfrac{(26.05 + 26.15)}{2}$ = 26.10 **(2)**

(iii) Moles of HCl in 26.10 cm³ of 0.100 mol/dm³ solution = 26.10 × 0.100/1000 = 0.002610 mol **(1)**

$HCl(aq) + KOH(aq) \rightarrow KCl(aq) + H_2O(l)$; ratio is 1 mol HCl : 1 mol KOH **(1)**

So amount of KOH reacting = 0.002610 moles KOH **(1)**

This is the amount of KOH in 25 cm³ solution in the flask, so 0.00261 = $25 \times \dfrac{conc}{1000}$ **(1)**

Therefore, concentration of KOH = 0.10 mol/dm³ **(1)**

36. Titration calculations

1 Moles of lithium hydroxide = volume in cm³ × $\dfrac{\text{concentration in mol/dm}^3}{1000}$ = 15.0 × 0.2/1000 = 0.003 **(1)**

1 mole of LiOH reacts with 1 mole of HNO_3 **(1)**

0.003 moles of LiOH reacts with 0.003 moles of HNO_3 **(1)**

0.003 × 1000/5.0 **(1)** = 0.6 mol/dm³ **(1)**

2 Moles of HCl = 0.0075, so moles of NaOH = 0.0075 **(1)**

0.0075 × 1000/25 **(1)**

= 0.3 mol/dm³ **(1)**

3 Moles of hydrochloric acid = 25.6 × 0.100/1000 = 0.00256 mol **(1)**

Moles of sodium carbonate = 0.00256/2 = 0.00128 mol **(1)**

0.00128 × 1000/25 **(1)**

= 0.0512 mol/dm³ = 0.051 mol/dm³ to two significant figures **(1)**

4 Moles of sulfuric acid = 1.0 × 20.0/1000 = 0.02

Using the ratio, moles of NaOH needed = 0.02 × 2 = 0.04

So, 20.0 cm³ of concentration 2.0 mol/dm³ is correct, because it is 0.04 moles. **(1)**

37. Reactions with gases

1 34/71 = 0.48 mol **(1)**

0.48 × 24 = 11.5 dm³ **(1)**

2 Moles HCl = 25 × 0.10/1000 **(1)** = 0.0025 **(1)**

Moles of H_2 = 0.00125 **(1)**

0.00125 × 24 = 0.03 dm³ **(1)**

3 (a) 300 cm³ **(1)**

(b) 400 cm³ **(1)**

4 (a) Moles of HCl = 20 × 0.2/1000 **(1)** = 0.004 **(1)**

Moles of CO_2 = 0.004/2 = 0.002 **(1)**

0.002 × 24 = 0.048 dm³ **(1)**

= 48 cm³ **(1)**

(b) Moles of calcium carbonate = 0.002 **(1)**

Mass of $CaCO_3$ = 0.002 × 100 (M_r) **(1)** = 0.2 g **(1)**

38. Reaction yields

1 (a) 0.8 g **(1)**

(b) 1.0 g **(1)**

(c) Some product probably escaped because there is no lid on the crucible.

It may not have been heated for long enough, so not all the magnesium had reacted. **(2)**

2 (a) Moles of copper carbonate = mass/relative formula mass = 10.0/123.5 **(1)** = 0.08 **(1)**

1 mole of copper carbonate produces 1 mole of copper nitrate.

Moles of copper nitrate = 0.08 **(1)**

Moles of copper nitrate × relative formula mass = mass of copper nitrate

0.08 × 187.5 **(1)** = 15 g **(1)**

(b) Percentage yield = 8/15 × 100 **(1)** = 53% **(1)**

3 69% **(1)**

39. Atom economy

1 (a) % atom economy = (relative formula mass of the moles of desired product in the equation)/(sum of formula masses of all the reactants in the equation) × 100 **(1)**

(b) 4/48 × 100 = 8.3% **(2)**

(c) If a use could be found for the carbon dioxide, then the atom economy would increase to 100%. **(1)**

2 (a) 28/142 × 100 = 19.7% **(2)**

(b) 100% **(1)**

(c) A reaction with a high atom economy is more economical because less material is wasted and so more of the raw materials bought can be sold on. There are fewer disposal costs for waste as there is less waste. The reduction in waste also makes the process more sustainable as it uses less raw material per finished product.

(2) (1 for each valid point made)

3 2 × 63.5/([2 × 63.5] + [2 × 16] + 12) × 100 = 74.3% **(1)**

40. Exam skills – Quantitative chemistry

1 (a) Moles of Ca = 0.23/40 **(1)** = 0.00575 **(1)**

Ratio is 1Ca : 1H_2 **(1)**

0.00575 Ca: 0.00575 H_2 moles **(1)**

0.00575 × 24 = 0.138 dm³ **(1)** = 0.14 dm³ to two significant figures **(1)**

(b) Ratio is 1Ca : 2HNO_3

Moles HNO_3 needed = 0.00575 × 2 = 0.0115 **(1)**

0.0115 × 1000/0.2 **(1)**

= 57.5 cm³ **(1)**

41. Reactivity series

1 zinc, lead, copper, silver **(1)**

2 (a) magnesium hydroxide + hydrogen **(1)**

(b) calcium nitrate + hydrogen **(1)**

(c) zinc chloride + hydrogen **(1)**

3 (a) hydrogen **(1)**

(b) hydroxide ion **(1)**

(c) The potassium atoms lose electrons (to form K⁺ ions). **(1)**

(d) Lithium is less reactive because it forms a positive ion less readily than potassium **(1)**. This is because the outer electron is closer to the nucleus and more tightly bound. **(1)**

4 zinc, iron, tin, lead, copper **(2)**

42. Oxidation

1 (a) H_2 **(1)**

(b) Calcium has gained oxygen **(1)**. Gain of oxygen is oxidation. **(1)**

(c) Calcium is a metal and loses electrons to form positive ions in this reaction **(1)**. Loss of electrons is oxidation. **(1)**

(d) Ca → Ca^{2+} + 2 e⁻ **(1)**

(e) Mg + Fe^{2+} → Mg^{2+} + Fe **(2)**

(f) magnesium **(1)**

2 (a) (i) white **(1)**

(ii) $2Mg + O_2 \rightarrow 2MgO$ **(1)**

(b) $C + O_2 \rightarrow CO_2$ **(1)**. Carbon has gained oxygen **(1)**. Gain of oxygen is oxidation. **(1)**

(c) $4KNO_3 \rightarrow 2K_2O + 2N_2 + 5O_2$ **(2)**

43. Reduction and metal extraction

1 (a) silver **(1)**

(b) Any two from: magnesium, tin, calcium **(2)**

(c) carbon **(1)**

(d) One from: calcium, magnesium **(1)**

2 (a) They are unreactive. **(1)**

(b) (i) In the reaction, CO has gained oxygen to form CO_2 **(1)**. Gain of oxygen is oxidation **(1)**. Fe_2O_3 has lost oxygen **(1)**; loss of oxygen is reduction. **(1)**

(ii) $Fe^{3+} + 3e^- \rightarrow Fe$ **(1)**

3 (a) $Mg + Cu^{2+} \rightarrow Mg^{2+} + Cu$ **(2)**

(b) sulfate ion **(1)**

(c) magnesium **(1)**

(d) $Mg \rightarrow Mg^{2+} + 2e^-$ **(1)**

44. Reactions of acids

1 (a) base **(1)**

(b) magnesium chloride, aluminium chloride **(2)**

2

Acid	Base	Salt
hydrochloric acid	lithium hydroxide	lithium chloride
nitric acid	calcium oxide	calcium nitrate
sulfuric acid	sodium hydroxide	**sodium sulfate**

(3)

3 (a) magnesium **(1)**

(b) potassium sulfate + water **(1)**

(c) sodium nitrate + water + carbon dioxide **(1)**

(d) copper sulfate + water **(1)**

4 (a) Add an indicator to the calcium hydroxide solution **(1)** and then add the acid slowly from a burette until the indicator changes colour. **(1)**

(b) $Ca(OH)_2 + 2HCl \rightarrow CaCl_2 + 2H_2O$ **(2)**

(c) Ca → Ca^{2+} + 2e⁻ **(1)**

The calcium has lost electrons **(1)**, and loss of electrons is oxidation. **(1)**

45. Core practical – Salt preparation

1 Add excess cobalt oxide to a measured quantity of dilute hydrochloric acid in a beaker and stir **(1)**. Filter the solution to remove excess cobalt oxide **(1)**. Heat the filtrate to evaporate some of the water, and crystals start to form **(1)**. Leave it to cool and crystallise. **(1)**

2 Group 1 would be unsuccessful **(1)** because copper metal does not react with sulfuric acid. **(1)**

(b)

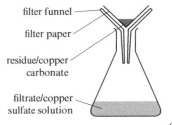

filter funnel

filter paper

residue/copper carbonate

filtrate/copper sulfate solution

(2)

(c) Heat the filtrate to evaporate some of the water, and crystals start to form **(1)**. Leave to cool and crystallise. **(1)**

(d) (i) sodium sulfate **(1)**

(e) (ii) Sodium is very reactive **(1)** so the reaction would be too dangerous/violent. **(1)**

46. The pH scale

1 (a) A and B **(1)**

(b) E **(1)**

(c) No change **(1)**

(d) The student could add some universal indicator **(1)** and compare the resulting colour with a colour chart to determine the pH. **(1)** (Alternatively a pH meter could be used.)

2 (a) $H_2SO_4 + 2KOH \rightarrow K_2SO_4 + 2H_2O$ **(2)**

(b) A strong acid is an acid that is completely/fully ionised **(1)** in aqueous solution **or** when dissolved in water. **(1)**

(c) $H^+(aq) + OH^-(aq) \rightarrow H_2O(l)$ **(2)**

(d) pH 5.1 **(1)**

3 (a) A **(1)**

(b) A is more concentrated as it contains 2 moles of acid in $1\,dm^3$; the nitric acid, B, contains only 0.5 moles of acid in $1\,dm^3$ solution. **(2)**

(c) Nitric acid is fully ionised into hydrogen ions in aqueous solution, as it is a strong acid, and so it has a low pH **(1)**. In contrast, ethanoic acid is a weak acid and is not completely ionised in solution, hence there are fewer hydrogen ions in an ethanoic acid solution of the same concentration as the nitric acid, and a higher pH. **(1)**

47. Electrolysis

1 (a) The lamp would light. **(1)**

(b) Solid lead bromide does not conduct electricity, as the ions cannot move and thus cannot carry charge **(1)**. When lead bromide is molten, the ions can move and carry charge. **(1)**

(c) anode: bromine **(1)**; cathode: lead **(1)**

(d) Reduction is gain of electrons **(1)**. At the anode, the bromine atoms gain electrons and form bromide ions. **(1)**

(e) It is a good conductor of electricity/does not react. **(1)**

2 (a) Metal ions are positively charged and move to the negative cathode **(1)**, where they discharge as they gain electrons/are reduced to the metal. **(1)**

(b) oxidation **(1)**

3

	Anode	Cathode
Product	chlorine **(1)**	sodium **(1)**
Half-equation	$2Cl^- \rightarrow Cl_2 + 2e^-$ **(2)**	$Na^+ + e^- \rightarrow Na$ **(2)**
Oxidation or reduction?	oxidation **(1)**	reduction **(1)**

48. Aluminium extraction

1 (a) The cryolite lowers the melting point **(1)**, so less energy is needed and the process is more economical. **(1)**

(b) Aluminium ions are positive and they move to the negative electrode **(1)**, where they gain electrons and are reduced, forming aluminium. **(1)**

(c) oxygen, carbon dioxide **(2)**

(d) $Al^{3+} + 3e^- \rightarrow Al$ **(2)**

(e) cathode **(1)**

(f) $2O^{2-} \rightarrow O_2 + 4e^-$ **(2)**

(g) Aluminium is a reactive metal and cannot be displaced from its ore by carbon. **(1)**

(h) Oxygen is formed at the anode **(1)**, where it reacts with the carbon of the electrode **(1)** to produce carbon dioxide. **(1)**

49. Electrolysis of solutions

1 It contains ions that can move. **(1)**

2 hydrogen, bromine **(1)**

3 (a) (i) hydrogen **(1)**

(ii) oxygen **(1)**

(iii) $2H^+ + 2e^- \rightarrow H_2$ **(2)**

(b) (i)

Electrolyte solution	Anode	Cathode
copper(II) chloride	chlorine	copper
potassium iodide	iodine	hydrogen
sodium bromide	bromine	hydrogen
sodium sulfate	oxygen	hydrogen

(4) (1 for each complete row)

(ii) An electrolyte is a molten ionic compound or a solution containing a dissolved ionic compound that can carry electricity. **(1)**

50. Core practical – Electrolysis

1 (a) A cathode, B anode; graphite for electrodes **(3)**

(b) There are dissolved ions in tap water, which might affect the results. **(1)**

(c)

Solution	Potassium chloride	Calcium nitrate	Sulfuric acid	Zinc bromide	Silver nitrate
Observations at cathode	colourless gas	colourless gas	colourless gas	grey solid	white solid
Observations at anode	greenish gas	colourless gas	colourless gas	orange solution	colourless gas
Test used for product at cathode	insert a burning lighted splint result – pop	insert a burning lighted splint result – pop	insert a burning lighted splint result – pop		
Test used for product at anode	universal indicator paper turns red and bleaches	relights a glowing splint **(1)**	relights a glowing splint	universal indicator paper turns red and bleaches	relights a glowing splint
Identity of product at cathode	hydrogen **(1)**	hydrogen **(1)**	hydrogen **(1)**	zinc	silver
Identity of product at anode	chlorine **(1)**	oxygen **(1)**	oxygen **(1)**	bromine **(1)**	oxygen **(1)**
Half-equation for reaction at cathode	$2H^+ + 2e^- \rightarrow H_2$ **(2)**			$Zn^{2+} + 2e^- \rightarrow Zn$ **(2)**	$Ag^+ + e^- \rightarrow Ag$ **(2)**
Half-equation for reaction at anode	$2Cl^- \rightarrow Cl_2 + 2e^-$ **(2)**	$4OH^- \rightarrow O_2 + 2H_2O + 4e^-$ **(2)**	$4OH^- \rightarrow O_2 + 2H_2O + 4e^-$ **(2)**	$2Br^- \rightarrow Br_2 + 2e^-$ **(2)**	

(d) Carry out the electrolysis in a fume cupboard **(1)**, because chlorine is toxic. **(1)**

51. Extended response – Chemical changes

*Answer could include the following points.

- Calcium chloride is formed in both reactions.
- Water is the other product of the calcium hydroxide with hydrochloric acid reaction.
- Hydrogen is the other product of the calcium with hydrochloric acid reaction.
- Observations for calcium hydroxide with hydrochloric acid include:
  ○ solution remains colourless.
- Observations for calcium and hydrochloric acid include:
  ○ bubbles/effervescence/gas produced
  ○ solid disappears
  ○ colourless solution formed.
- Equations for the reactions are:
  ○ $Ca(OH)_2 + 2HCl \rightarrow CaCl_2 + 2H_2O$
  ○ $Ca + 2HCl \rightarrow CaCl_2 + H_2$

This answer is not exhaustive; other creditworthy responses will be awarded marks too.

52. Exothermic reactions

1 The amount of energy in the Universe at the end of a chemical reaction is the same as before the reaction takes place. **(2)**

2 (a) An exothermic reaction is one that transfers energy to the surroundings **(1)**, so the temperature of the surroundings increases. **(1)**

(b) The product molecules have less energy than the reactants (because energy is released to the surroundings during an exothermic reactions). **(1)**

(c) handwarmers, self-heating cans or other valid examples **(2)**

3 (a) $CH_4 + 2O_2 \rightarrow CO_2 + 2H_2O$ **(2)**

(b) increase

(c) and (d)

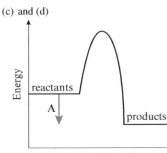

(c) labelled reactions and products **(1)**, reactants higher than products **(1)**, x-axis **(1)**, y-axis **(1)**

(d) labelled line in correct direction **(1)**

53. Endothermic reactions

1 (a) An exothermic reaction gives out **(1)** heat energy while an endothermic reaction takes in **(1)** heat energy.

(b) Place a thermometer in the solution **(1)** and observe a temperature drop. **(1)**

(c) One from: thermal decomposition, citric acid with sodium hydrogen carbonate, photosynthesis **(1)**

2 The equation for the combustion of glucose with oxygen states that energy is given out in the reaction **(1)**, hence energy must be taken in in the reverse reaction (photosynthesis), which means it is endothermic. **(1)**

3 (a) A exothermic, B endothermic, C exothermic, D endothermic **(4)**

(b) A +29, D −11 **(2)**

54. Core practical – Energy changes

1 (a) It is a poor conductor of heat, so no energy is lost to the surroundings. **(1)**

(b) to ensure the two solutions mixed completely and fully reacted **(1)**

(c) Add a lid **(1)** to reduce heat loss. **(1)**

(d) $25 \times 1/1000 = 0.025$ mol **(1)**

(e) (i) exothermic **(1)**, because the temperature increased in each experiment **(1)**

(ii) The experiment was carried out using the same volume and concentration **(1)** of acid and the same volume and concentration **(1)** of alkali, and was started at the same temperature. **(1)**

(iii) Ethanoic acid is a weak acid **(1)**, so it does not ionise completely **(1)**.

This means there are fewer H^+ ions available to react and less energy is released. **(1)**

(iv) Add an indicator/use a pH meter **(1)** and repeat the experiment. Record the temperature at the colour change/pH 7 and see if it coincides with the maximum temperature. **(1)**

55. Activation energy

1 (a) (i) 2 **(1)**

(ii) 3 **(1)**

(b) The products have less energy than the reactants so energy has been given out to the surroundings. **(2)**

(c) 50 kJ **(1)**

2 (a) $50 - 20 = 30$ kJ **(1)**

(b) $100 - 20 = 80$ kJ **(1)**

(c) the minimum amount of energy needed for particles to react **(1)**

3 Reaction 1 endothermic **(1)**, Reaction 2 exothermic **(1)**, Reaction 3 exothermic **(1)**

56. Bond energies

1 (a) Energy in: $941 + 1308 = 2249$ kJ. Energy out: $6 \times 391 = 2346$ kJ. Energy change = -97 kJ **(3)**

(b) The reaction is exothermic **(1)** because more energy is released when the new bonds are made than is taken in to break existing bonds. **(1)**

2 Energy is needed to break the bonds **(1)** and overcome the activation energy barrier. **(1)**

3 (a) Bonds broken = $436 + 193 = 629$ kJ **(2)**

(b) $629 - 2x = -103$ **(1)**

$x = $ bond energy $= (629 + 103)/2 = 366$ kJ **(1)**

57. Cells

1 (a) $2H_2 + O_2 \rightarrow 2H_2O$ **(2)**

(b) $H_2 \rightarrow 2H^+ + 2e^-$ **(1)**

$O_2 + 4H^+ + 4e^- \rightarrow 2H_2O$ **(1)**

(c) hydrogen fuel cell: The only waste product is water; rechargeable cell: portable **(2)**

2 (a) B, A and D are more reactive than copper because the voltage observed is positive for B, A and D **(1)**, with B more reactive than A, then D. C is less reactive than copper **(1)**, so the order is: B, A, D, copper, C. **(1)**

(b) 0 V **(1)**

(c) the electrolyte or concentration of the electrolyte **(1)**

(d) ammonium chloride solution **(1)**

58. Extended response – Energy changes

$H_2 + Cl_2 \rightarrow 2HCl$

Energy in = $436 + 243 = 679$ kJ/mol

Energy out = $2 \times 432 = 864$ kJ/mol

Energy change = in − out = $679 - 864 = -185$ kJ/mol **(2)**

$H_2 + Br_2 \rightarrow 2HBr$

Energy in = $436 + 193 = 629$ kJ/mol

Energy out = $2 \times 366 = 732$ kJ/mol

Energy change = in − out = $629 - 732 = -103$ kJ/mol **(2)**

Both energy changes are negative **(1)**, showing that energy is released to the surroundings and both reactions are exothermic. **(1)**

This answer is not exhaustive; other creditworthy responses will be awarded marks too.

59. Rate of reaction

1 (a) (i) Rate = change/time = $(0.36 - 0.22)/(4 - 2) = 0.07$ g/min **(2)**

(ii) Rate = $(0.42 - 0.36)/(6 - 4) = 0.03$ g/min **(2)**

(b) Time taken = $2 \times 60 = 120$ seconds

Rate = change/time = $(0.30 - 0.12)/120 = 0.0015$ g/s **(3)**

(c) at 7 minutes **(1)**, because there is no further loss in mass **(1)**

(d) $0.45/(7 \times 60)$ **(2)** = $0.45/420 = 0.001071$ **(1)** = 0.0011 g/s to two significant figures **(1)**

(e) conical flask on balance **(1)**, acid and calcium carbonate in flask **(1)**, cotton wool **(1)**

60. Rate of reaction on a graph

1 (a) $Mg + 2HCl \rightarrow MgCl_2 + H_2$ **(2)**

(b) axes labelled with sensible scales **(1)**, all points from table plotted taking up more than half the grid **(1)** and a smooth curve drawn **(1)**

2 (a) $30/20 = 1.5$ cm^3/s **(2)**

(b) $15/20 = 0.75$ cm^3/s **(2)**

(c) 70 seconds **(1)**

(d) result at 30 seconds in experiment A **(1)**

61. Calculating the gradient

1 (a) The result at 120 seconds does not fit on the curve and is anomalous. **(1)**

(b) The reaction is complete at 130 seconds. **(1)**

Mean rate = mass loss/time **(1)** = $(126.0 - 125.78)/130$ **(1)** = $0.22/130 = 0.0017$ **(1)** g/s **(1)**

(c) Draw a tangent to the curve at 100 seconds **(1)** and find the gradient of the tangent. **(1)**

2 Draw a tangent to the curve at 10 seconds. **(1)** Gradient = change in y-values/change in x-values = $52/30 = 1.73$ **(1)** cm^3/s **(1)**

62. Collision theory

1 when particles collide with sufficient energy **(1)**

2 (a) As the reaction proceeds, the reactant particles are used up **(1)** and their concentration decreases **(1)**. Therefore collisions between reactant particles are less frequent/there are fewer collisions per second **(1)** and the rate decreases/reaction slows down. **(1)**

(b) In the second experiment, because the acid was more concentrated, there were more particles **(1)** in the same volume **(1)**. This means there were more frequent collisions/more collisions per second **(1)** and thus a faster rate of reaction would have been observed. **(1)**

3 (a) $H_2 + Br_2 \rightarrow 2HBr$ **(2)**

(b) the minimum amount of energy needed for particles to react **(1)**

Answers

63. Rate: pressure, surface area

1 calcium carbonate powder reacting with an excess of $2\,mol/dm^3$ nitric acid **(1)**

2 (a) $70\,s$ **(1)**

(b) The surface-area-to-volume ratio was greater for the powder than for the ribbon **(1)**, so the solid had a larger exposed surface for the other reactant particles to collide with **(1)**. There would have been more frequent collisions **(1)** so the rate of reaction increased. **(1)**

(c) line A **(1)**: it has a steeper slope, showing a faster reaction, but the same volume of gas is produced **(1)**

(d)

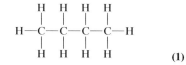

gas syringe

conical flask

magnesium and acid

stopwatch

(4)

64. Rate: temperature

1 (a) (i) Flask A **(1)**

(ii) Any two from: least reactive metal, lowest temperature, most dilute acid, smallest surface-area-to-volume ratio **(2)**

(b) (i) Flasks E and F **(1)**

(ii) Any two from: same volume of acid, same mass of magnesium, same surface-area-to-volume ratio **(2)**

2 (a) Increasing the temperature makes the reaction rate increase. **(1)**

(b) The higher the temperature, the higher the proportion of particles that have more energy than the activation energy required **(1)**. Reactant particles with more energy move faster **(1)**, so the frequency of collisions is increased **(1)** at a higher temperature.

65. Core practical – Rate of reaction

1 (a) stopwatch **(1)**

(b) so that none of the gas escapes and it is all collected in the gas syringe **(2)**

(c) bubbles, metal will disappear **(2)**

(d) Any two from: mass of magnesium, volume of acid, form of magnesium (ribbon only), temperature **(2)**

(e) Increasing the concentration means the slope of the graph becomes steeper **(1)** and the reaction rate is higher (reaction is faster). **(1)**

(f) There were more particles present in the same volume, so there were more collisions per second/more frequent collisions of particles **(1)** that have more energy than the activation energy. **(1)**

66. Catalysts

1 Graph C **(1)**: a catalyst is not used up so the mass is constant **(1)**

2 (a) gas syringe **(1)**

(b) A catalyst increases the rate of reaction by providing a different pathway for the reaction **(1)** that has a lower activation energy. **(1)**

(c) (i) $48\,cm^3$ **(1)**

(ii) zinc oxide **(1)**, because the reaction with ZnO catalyst took the most time **(1)**

(iii) The line should start at (0,0) and remain steeper and above the graph line, but level off earlier, at the same volume. **(1)**

67. Reversible reactions

1 a reaction where the products of the reaction can react to produce the original reactants **(1)**

2 (a) $N_2(g) + 3H_2(g) \rightleftharpoons 2NH_3(g)$ **(2)**

(b) $\rightleftharpoons$ double arrow **(1)**

3 (a) This means the reaction goes in both directions. **(1)**

(b) carbon monoxide, hydrogen, methane, water **(2)**

4 (a) the reverse direction (remember, heat drives water off hydrated copper sulfate) **(1)**

(b) white to blue **(1)**

(c) contains water **(1)**

68. Equilibrium and Le Chatelier's principle

1 (a) At equilibrium, the forward and reverse reactions **(1)** occur at exactly the same rate. **(1)**

(b) One from:
  - The apparatus must be sealed/prevent the escape of reactants and products.
  - A closed system must be used. **(1)**

(c) The position of equilibrium will move to the right **(1)**, which increases the pressure **(1)** as there are more molecules on that side of the equation **(1)**. Therefore, more hydrogen will form. **(1)**

2 (a) advantage: increase in % yield of ammonia **(1)**; disadvantage: costs increase **(1)** (owing to the need for high-pressure equipment and increased safety precautions)

(b) (i) higher yields **(1)**

(ii) The rate of reaction becomes too slow to be commercially useful. **(1)**

69. Equilibrium: changing temperature and pressure

1 (a) (i) a reaction that takes in heat **(1)**

(ii) a reaction that occurs in both directions **(1)**

(b) If the temperature is raised, the yield in the endothermic direction will increase **(1)**, so there will be more NO_2 **(1)** and the colour will change to brown. **(1)**

(c) There are two molecules on the right and one on the left **(1)**. If the gas pressure is increased, the equilibrium moves to reduce the pressure **(1)**, so it would move left and the colour would turn more yellow. **(1)**

2 (a) There are two moles on each side of the equation so the equilibrium position is unchanged **(1)** and the amount of product is unchanged. **(1)**

(b) There are three moles on the left and two on the right, so decreasing the pressure means the equilibrium position moves left **(1)** and the amount of the products decreases. **(1)**

3 The formation of methanol is an exothermic reaction, so lowering the temperature would mean that the equilibrium moves in the direction that gives out heat, to the right, giving more product. **(1)** There are fewer moles of gas on the right, so increasing the pressure would mean that the equilibrium moves to the right, giving more product. **(1)**

70. Extended response – Rate of reaction and equilibria

*Answer could include the following points.

- Plan includes:
  ○ add magnesium to acid
  ○ time reaction/measure volume of gas (with method)
  ○ change concentration and repeat.
- Fair test:
  ○ same mass of magnesium
  ○ same surface area of magnesium
  ○ same volume of acid
  ○ same temperature.

This indicative content is not exhaustive; other creditworthy responses will be awarded marks too.

71. Crude oil

1 (a) C_{11}–C_{13} **(1)**

(b) (i) residue above C_{20} **(1)**

(ii) petrol **(1)**

(c) fractional distillation **(1)**

2 (a) The crude oil is heated until most of it has evaporated **(1)**. It then passes into a fractionating column **(1)**. As the mixture rises up the column **(1)**, the gases cool and condense at different temperatures. **(1)**

(b) (i) Any two from: fuel oil, kerosene, diesel, gasoline **(2)**

(ii) Any three from: solvents, lubricants, polymers, detergents **(3)**

72. Alkanes

1 (a) a compound made up of hydrogen **(1)** and carbon atoms only **(1)**

(b)

Alkanes	Formula	Boiling point in °C
methane	CH_4	−162
ethane	C_2H_6	−89
propane	$\mathbf{C_3H_8}$	**−40**
butane	C_4H_{10}	0
pentane	C_5H_{12}	+36

(3)

2 (a)

(1)

130

(b)

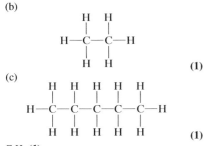

(c)

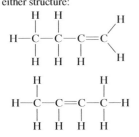

(1)

3 C_4H_8 **(1)**

73. Properties of hydrocarbons

1 (a) In tube A, a colourless liquid **(1)** would be observed. In tube B, the limewater would change from colourless to cloudy. **(1)**

(b) Anhydrous copper sulfate **(1)** changes from white **(1)** to blue **(1)** (in the presence of water).

c) hydrocarbon + oxygen → carbon dioxide + water **(1)**

(d) $C_4H_{10} + 6.5O_2 \rightarrow 4CO_2 + 5H_2O$ **(2)**

2 (a) octane: it has larger molecules **(2)**

(b) pentane **(1)**

(c) so that the fuel flows easily (and can be transported in and through the engine easily) **(1)** and catches fire easily (so burning in the engine is straightforward and controllable) **(1)**

74. Cracking

1 (a) breaking up a large molecule to produce a smaller, more useful molecule **(2)**

(b) Pass the vapour over a hot catalyst or mix it with steam and heat to high temperature. **(2)**

(c) $C_8H_{18} \rightarrow C_2H_4 + C_6H_{14}$ **(2)**

(d) Smaller hydrocarbons make better fuels than larger ones. **(1)** Alkenes can be used to make polymers. **(1)**

2 thermal decomposition **(1)**

3 (a) Crude oil B is the most viscous as it has the largest proportion of large molecules, such as bitumen. **(2)**

(b) The percentages of fuel oil and bitumen would decrease, as the large molecules are broken into smaller molecules. **(2)**

75. Alkenes

1 (a) B and D **(1)**

(b) C propane, D ethene **(2)**

(c) They all contain hydrogen and carbon only. **(2)**

(d) (i) $C_3H_8 + 5O_2 \rightarrow 3CO_2 + 4H_2O$ **(2)**

(ii) $C_2H_4 + 3O_2 \rightarrow 2CO_2 + 2H_2O$ **(2)**

(e) C_nH_{2n}

2 (a) $x = 4, y = 8$ **(2)**

(b) butene **(1)**

(c) either structure:

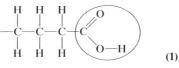

(1)

76. Reactions of alkenes

1

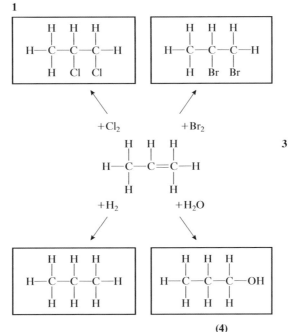

(4)

2 (a) C=C **(1)**

(b) a smoky flame **(1)** due to incomplete combustion as the C=C bond is difficult to break **(1)**

(c) steam/water, catalyst **(2)**

(d) chlorine **(1)**

(e) hydrogen, catalyst **(2)**

(f) Shake with bromine water **(1)**; the mixture would change from orange to colourless with A **(1)**, but would stay orange with D. **(1)**

77. Alcohols

1 (a) OH/hydroxyl **(1)**

(b) C_4H_9OH **(1)**

2 (a) carbon dioxide, water **(2)**

(b) Butanol could be used as a solvent or as a fuel. **(2)**

(c) $C_4H_9OH + 6O_2 \rightarrow 4CO_2 + 5H_2O$ **(2)**

3 (a) CH_3OH

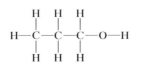

(2)

(b) Both these alcohols react with **sodium** metal to produce **hydrogen** gas. **(2)**

4

(1)

5 Any three from: yeast, warm temperature, aqueous solution, no air **(3)**

78. Carboxylic acids

1 (a) C_3H_7COOH **(1)**

(b)

(1)

2 (a) butanol **(1)**

(b) C_2H_5COOH

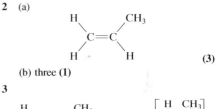

(2)

(c) carbon dioxide **(1)**

(d) Carboxylic acids react with alcohols, forming esters/an ester. **(2)**

3 (a) Formula: CH_3COOH

pH: approx. 3 or 4

Time taken: approx. 1 or 2 minutes **(3)**

(b) Weak acids do not ionise completely in solution so there are fewer H^+ ions present **(1)**. Strong acids do ionise completely in solution. **(1)**

(c) Hydrochloric acid ionises completely in solution **(1)**, giving a higher concentration of H^+ ions and so a lower pH **(1)**. Ethanoic acid does not ionise completely and so there is a lower concentration of H^+ ions and a higher pH. **(1)**

79. Addition polymers

1 (a) ethene, poly(ethene) **(2)**

(b) Polymers are long chain molecules made by joining many monomers in a polymerisation reaction. **(2)**

2 (a)

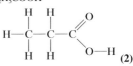

(3)

(b) three **(1)**

3

(2)

4

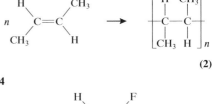

(1)

80. Condensation polymers

1 polyester **(1)**

2 (a) Condensation polymerisation occurs when monomers with two functional groups react and join together, usually losing small molecules such as water. **(2)**

(b) Condensation polymers are formed from monomers with two functional groups, but addition polymers are formed from monomers with one functional group/multiple bond. In condensation polymerisation, the product is the polymer and a small molecule such as water, but in addition polymerisation the only product is the polymer. **(2)**

(c)

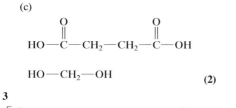

(2)

3

(2)

81. Biochemistry

1 (a) A **(1)**

(b) A: C=C **(1)**, B: COOH **(1)** and NH₂ (H₂N) **(1)**, C: COOH **(1)**, D: OH **(1)**

(c) B **(1)**

(d) C ethanoic acid **(1)**, D ethanol **(1)**

(e) B **(1)** (it is an amino acid and forms a protein)

2 (a) DNA **(1)**

(b) double helix **(1)**

(c) (i) 2 **(1)**

(ii) 4 **(1)**

82. Extended response – Organic chemistry

*Answer could include the following points.

Fermentation

- Start with sugar and add yeast.
- Produces carbon dioxide and ethanol.
- Conditions: warm temperature, no air (anaerobic), aqueous solution.

Distillation

- Heat the dilute ethanol.
- Use antibump for even boiling.
- The ethanol boils first and vapour moves into the condenser.
- Vapour cools and condenses, and is collected.

This answer is not exhaustive; other creditworthy responses will be awarded marks too.

83. Pure substances and formulations

1 Determine the melting point. **(1)**

2 (a) It is a mixture of iron and carbon, which are added in carefully measured quantities to ensure that the mixture has the required properties for a particular purpose. **(2)**

(b) Any two from: medicines, fuels, cleaning agents, paints, fertilisers **(2)**

(c) (i) A solid, B solid, C liquid **(3)**

(ii) A element (it has a sharp melting point and boiling point), B formulation (melting and boiling points are not sharp but have ranges), C element (sharp melting point and boiling point) **(3)**

3 (a) Compare the measured melting point with those included in a data book. **(1)**

(b) The melting point of an element would be sharp, not a range. **(1)**

84. Core practical – Chromatography

1 (a) A (purple food) dye **(1)**, B filter paper/ chromatography paper **(1)**, C beaker **(1)**, D solvent **(1)**

(b) The paper is not touching the solvent **(1)**, so the solvent cannot move up the paper and over the dye to separate the constituents of the dye. **(1)**

2 (a) 1 **(1)**

(b) so that it does not dissolve in the solvent **(1)**

(c) No. It separated into three different spots, which each moved a different distance up the paper, showing that it contains three different substances. **(1)**

(d) R_f = distance moved by substance/ distance moved by solvent = 1.9/3.7 = 5.1 **(2)**

(e) The mobile phase/solvent moves upwards through the paper and carries different compounds for different distances that depend on their attraction for paper and solubility in the solvent. **(3)**

85. Tests for gases

1 oxygen **(1)**

2 (a) $CaCO_3 + 2HCl \rightarrow CaCl_2 + H_2O + CO_2$ **(2)**

(b) Solution changed from colourless to milky **(1)**, because carbon dioxide produced in the reaction of acid and carbonate **(1)** reacted with the limewater. **(1)**

(c) calcium hydroxide; $Ca(OH)_2$ **(2)**

3 Add a burning splint – hydrogen pops with a lighted splint. **(1)**

Add a glowing splint – oxygen relights the splint. **(1)**

Add a piece of damp litmus paper – it turns red and then is bleached/turns white if the gas is chlorine. **(1)**

Helium – no reaction with any of above. **(1)**

86. Tests for cations

1 iron(II) sulfate solution **(1)**

2 (a) Dip a flame test wire loop in acid then heat in a Bunsen flame until there is no colour **(1)**. Dip the loop in acid again and then into the compound. Hold the loop in the Bunsen flame and record the colour. **(1)**

(b) Ionic substances contain two ions **(1)** and both must be identified. **(1)**

(c) carbon dioxide **(1)**

(d) calcium carbonate **(2)**

3 (a) precipitation **(1)**

(b) Cu²⁺ (aq), green solid formed (Fe²⁺), brown solid formed (Fe²⁺) **(3)**

(c) By adding excess sodium hydroxide solution, both aluminium and magnesium ions will produce a white precipitate. The precipitate for aluminium will then redissolve to give a colourless solution. **(2)**

(d) $Ca(OH)_2$ **(1)**; (s) **(1)**; (aq) **(1)**

87. Tests for anions

1 First add hydrochloric acid. If carbonate ions are present a gas is produced that will turn limewater cloudy **(1)**. To test for sulfate ions, add barium chloride solution and, if they are present, a white precipitate will form. **(1)**

2 (a) Dissolve each solid in distilled water to make test solutions. Make up a solution of silver nitrate by adding silver nitrate solid to nitric acid solution **(1)**. Add a few drops of silver nitrate solution to the test solutions **(1)**. The formation of a white precipitate indicates chloride, while a cream precipitate indicates bromide. **(1)**

(b) Dissolve each solid in distilled water to make test solutions. Make up a solution of silver nitrate by adding silver nitrate solid to nitric acid solution **(1)**. Add a few drops of silver nitrate solution to the test solutions **(1)**. The formation of a white precipitate indicates chloride. **(1)**

or

Dissolve each solid in distilled water to make test solutions. Make up a solution of barium chloride by adding barium chloride solid to distilled water **(1)**. Add a few drops of barium chloride solution to the test solutions **(1)**. The formation of a white precipitate indicates sulfate. **(1)**

3 (a) Test 1: it does not contain sodium, potassium, calcium, copper, or lithium ions. **(1)**

Test 2: it contains aluminium ions. **(1)**

(b) Test 3: it is not a carbonate. **(1)**

Test 4: it contains iodide ions. **(1)**

88. Flame emission spectroscopy

1 Carry out flame emission spectroscopy. **(1)**

2 (a) copper and calcium ions **(2)**

(b) A flame test is suitable as the two ions give different colours **(1)** (though it may be hard to distinguish the colours in one flame). Adding sodium hydroxide solution is also suitable as the two ions give different colours of precipitate. **(1)**

(c) Flame emission spectroscopy is (any two of) more accurate, more sensitive, can be used with small samples, rapid. **(2)**

(d) 2,8,8 **(1)**

89. Core practical – Identifying a compound

1 Dissolve the sample in deionised water **(1)**.

Add a few drops of sodium hydroxide solution, followed by an excess **(1)**.

If a white precipitate occurs that is soluble in excess NaOH, then the salt is an aluminium compound (or: if a white precipitate forms that is insoluble in excess NaOH, then it is a magnesium compound) **(1)**.

Add a few drops of nitric acid and then silver nitrate solution **(1)**.

The formation of a white precipitate indicates chloride, while a cream precipitate indicates bromide, and a yellow precipitate indicates iodide. **(1)**

2 (a) sodium **(1)**

(b) aluminium, Al^{3+} **(2)**

(c) I^- **(1)**

(d) aluminium iodide, sodium iodide **(2)**

(e) $Ag^+(aq) + I^-(aq) \rightarrow AgI(s)$ **(2)**

90. Extended response – Chemical analysis

*Answer could include the following points.

Magnesium sulfate solution

- Add sodium hydroxide solution – forms a white precipitate, which does not dissolve on adding excess NaOH.
- Add barium chloride solution – forms a white precipitate.

Sodium chloride solution

- Flame test – gives a yellow flame.
- Add nitric acid and silver nitrate solution – forms a white precipitate.

Iron(II) iodide solution

- Add sodium hydroxide solution – forms a green precipitate, which does not dissolve on adding excess NaOH.
- Add nitric acid and silver nitrate solution – forms a yellow precipitate.

Magnesium bromide solution

- Add sodium hydroxide solution – forms a white precipitate, which does not dissolve on adding excess NaOH.
- Add nitric acid and silver nitrate – forms a cream precipitate.

This answer is not exhaustive; other creditworthy responses will be awarded marks too.

91. The early atmosphere and today's atmosphere

1 nitrogen **(1)**

2 (a) nitrogen 80, oxygen 20 **(2)**

(b) argon **(1)**

(c) methane, ammonia **(2)**

3 (a) Gas volume changes with temperature. **(1)**

(b) $2Mg + O_2 \rightarrow 2MgO$ **(2)**

(c) Volume of oxygen = 28 cm^3
28/200 × 100 = 14% **(2)**

(d) The oxygen was not all used up **(1)** because:

- not all of the magnesium reacted with it; or
- some of the magnesium may have reacted with nitrogen in the air instead. **(1)**

92. Evolution of the atmosphere

1 (a) Any three from:

- The early atmosphere contained no oxygen, whereas today's atmosphere contains 21% oxygen.

- The early atmosphere was mainly carbon dioxide (95.5%), but today's only has 0.04% carbon dioxide.
- The early atmosphere had only 3.1% nitrogen, but today's has 78% nitrogen.
- The Earth's atmosphere today has slightly less argon (0.9%) than the 1.2% present in the early atmosphere. **(3)**

(b) We can measure the modern atmosphere with accurate methods, including direct instrumental methods, whereas we have to use indirect methods for the early atmosphere. **(1)**

(c) Algae and plants decreased the percentage of carbon dioxide and increased the percentage of oxygen in the atmosphere **(1)** by photosynthesis **(1)**.

2 (a) CO_2 (g) **(1)**

(b) (i) shells **(1)**

(ii) limestone **(1)**

(c) $6CO_2 + 6H_2O \rightarrow C_6H_{12}O_6 + 6O_2$ **(1)**

93. Greenhouse gases

1 (a) water vapour **(1)**, which has the formula H_2O **(1)**, and methane **(1)**, with the formula CH_4 **(1)**

(b) Carbon dioxide allows short-wavelength radiation from the Sun to pass through the atmosphere to the Earth's surface **(1)**, but absorbs outgoing long-wavelength radiation **(1)**, preventing the Earth from cooling down. **(1)**

(c) The percentage of carbon dioxide decreased rapidly from 4500 to 3500 million years ago **(1)**.
The decrease then continued, but more gradually. **(1)**

(d) combustion of fossil fuels, deforestation **(2)**

94. Global climate change

1 (a) During photosynthesis, plants take in carbon dioxide and give out oxygen **(1)**, so reducing the number of plants can increase the levels of carbon dioxide. **(1)**

(b) Burning fossil fuels **(1)** releases carbon dioxide **(1)** into the atmosphere.

(c) Warming, caused by increased levels of carbon dioxide, results in (any two from):

- sea level rise, which may cause flooding and increased coastal erosion
- more frequent and severe storms
- changes in the amount, timing and distribution of rainfall
- temperature and water stress for humans and wildlife
- changes in the food-producing capacity of some regions
- changes to the distribution of wildlife species. **(2)**

2 (a) When the carbon dioxide concentration increased **(1)**, the temperature change increased **(1)**.

(b) The percentage of carbon dioxide is increasing **(1)**, causing global temperatures to increase. **(1)**

3 (a) an increase in the average global temperature **(1)**

(b) Any two from: sea level rise, flooding, change in weather and rainfall, change to distribution of wildlife species **(2)**

95. Carbon footprint

1 the total amount of all greenhouse gases emitted over the full life cycle of a substance **(1)**

2 (a) approximately 44% (accept 41–45%) **(1)**

(b) Any two from: use a bike, car share, use public transport, holiday locally **(2)**

(c) (i) Any two from: solar, hydroelectric, geothermal, windpower **(2)**

(ii) The carbon dioxide generated from power stations **(1)** could be collected and stored underground in used oil fields. **(1)**

(d) Any two from: they may not think that carbon dioxide causes global climate change, lack of public information and education, lifestyle changes are difficult or costly to make, economic considerations (e.g. train tickets for large family more expensive than trip by car) **(2)**

96. Atmospheric pollution

1 (a) carbon monoxide and soot **(1)**

(b) carbon – global dimming **(1)**, carbon monoxide – prevents the blood from taking up oxygen **(1)**, carbon dioxide – global warming **(1)**

2 (a) nitrogen from air, sulfur dioxide from the combustion of sulfur impurities in fossil fuels **(2)**

(b) 2% **(1)**

(c) Sulfur dioxide can cause breathing problems **(1)** and can contribute to acid rain. **(1)**

3 (a) When the petrol burns, the sulfur oxidises and sulfur dioxide is produced. **(1)**

(b) Oxides of nitrogen are produced by the reaction of nitrogen and oxygen from the air at the high temperatures involved when fuels are burned. **(1)**

(c) Carbon monoxide is produced by incomplete combustion of the fuel. **(1)**

(d) Soot (carbon particles) is produced by incomplete combustion of the fuel. **(1)**

97. Extended response – The atmosphere

*Answer could include the following points.

- The concentration of carbon dioxide was fairly steady up until 1800.
- The concentration of carbon dioxide increased dramatically from 1800 to 2000.
- This increase could be due to increased human activity, such as increased combustion of fossil fuels and deforestation.
- Many scientists believe that increased carbon dioxide will cause the temperature of the Earth's atmosphere to increase at the surface.
- This will result in global climate change.
- Effects of global climate change, e.g. sea level rise, flooding, change in weather and rainfall, change to distribution of wildlife species.

This answer is not exhaustive; other creditworthy responses will be awarded marks too.

98. The Earth's resources

1 renewable **(1)**

2 oil **(1)**

3 biodiesel: renewable, coal: finite, ethanol: renewable, wind power: renewable, petrol: finite **(5)**

4 Using natural gas is not sustainable (it compromises the ability of future generations to meet their own needs), as it uses up a finite resource and cannot go on forever **(1)**. Wind power, solar power or tidal power can be used to generate electricity in a sustainable way. **(1)**

5 wool: natural **(1)**

plastic: synthetic **(1)**

cotton: natural **(1)**

wood: natural **(1)**

99. Water

1 (a) It contains dissolved minerals. **(1)**

(b) Step 1: Pass it through a filter bed **(1)** to remove any solids. **(1)**

Step 2: Sterilise it using chlorine or ozone **(1)** to kill microbes. **(1)**

(c) (i) the removal of salt from water **(1)**

(ii) distillation, reverse osmosis **(2)**

(iii) It requires large amounts of energy. **(1)**

2 (a) to remove large solids and grit from the wastewater **(1)**

(b) effluent **(1)**

(c) sludge **(1)**

(d) Microbes break down **(1)** dissolved organic materials in the presence of air. **(1)**

100. Core practical – Analysis and purification of water

1 (a) in the conical flask **(1)**

(b) in the test tube **(1)**

(c) to cool the vapour and cause condensation **(1)**

(d) A conical flask, B delivery tube **(1)**

(e) Liebig condenser **(1)**

(f) Measure its boiling point; it should be 100°C. **(2)**

(g) Add barium chloride solution; if sulfate ions are present, a white precipitate will form. **(2)**

(h) It needs large amounts of energy and often this uses up finite resources (e.g. crude oil); using fossil fuels to provide energy causes carbon dioxide and other pollutants to be released, whch can lead to global warming or acid rain. **(2)**

101. Alternative methods of extracting metals

1 (a) Iron is more reactive than copper, so it displaces the copper from copper chloride solution. **(2)**

(b) iron + copper chloride → copper + iron(II) chloride **(1)**

(c) displacement / redox **(1)**

(d) zinc chloride **(1)**

(e) One from: more scrap iron available than scrap zinc, it is more profitable to recycle zinc, scrap iron is cheaper **(1)**

2 (a) The higher-grade ores have all been used up. **(1)**

(b) Bioleaching: some bacteria absorb metal compounds to produce a solution called a leachate. By adding a scrap metal that is more reactive than the metal in the leachate, the metal of interest can be displaced from solution. **(2)**

Phytomining uses plants to absorb metal compounds. The plants are harvested and then burned to produce ash that contains the metal compounds. **(2)**

(c) (i) One from: less energy is needed than traditional means, low-grade ores can be used, no release of sulfur dioxide, in phytomining the plants release energy when burned **(1)**

(ii) One from: it takes time to grow the plants, large volumes of leachate must be safely disposed of **(1)**

102. Life cycle assessment

1 Life cycle assessments are carried out to assess the total impact on the environment of a product over its whole life, from extracting the raw materials to its disposal. **(2)**

2 (a) 7.52×10^6 **(1)**

(b) Any three from:
- extracting and processing raw materials
- manufacturing and packaging
- use and operation during the product lifetime
- disposal at the end of its useful life
- transport and distribution at each stage **(3)**

(c) To manufacture once, and fill the glass bottle four times, the energy is
$7\,520\,000 + 2\,000\,000 \times 4 + 500\,000 \times 4$
$= 1.752 \times 10^7$ joules

To manufacture four new plastic bottles the energy = $4 \times$ manufacture + $4 \times$ filling
$= 4 \times 2\,200\,000 + 4 \times 4\,500\,000$
$= 2.68 \times 10^7$ joules

The energy saving =
$2.68 \times 10^7 - 1.752 \times 10^7 = 9\,280\,000$ J **(3)**

(d) Any two from:
- increased use of alternative energy supplies to manufacture the bottle
- energy conservation
- carbon capture and storage
- carbon offsetting, including through tree planting
- recycling. **(2)**

103. Conserving resources

1 (a) lead and aluminium **(1)**

(b) 1. Because its ore might be scarce. 2. Because it might be expensive to extract the metal. **(2)**

(c) 72% is recycled, so 72/100 × 4.6 = 3.312 = 3.3 million tonnes **(2)**

(d) reduces the use of limited resources, reduces energy consumption, reduces waste **(3)**

2 (a) $(27 \times 2)/(27 \times 2 + 16 \times 3) \times 100$
$= 52.9\%$ **(2)**

(b) (i) Aluminium can be recycled by melting and recasting or reforming into different products. **(2)**

(ii) It reduces the use of resources like bauxite, with the associated costs and energy usage of mining. It requires much less energy than to make something new, and reduces waste and the associated environmental impacts. **(2)**

104. Corrosion

1 (a) Corrosion is the destruction of materials by chemical reactions with substances in the environment. **(1)**

(b) air and water **(2)**

(c) Aluminium has an oxide coating that protects the metal from further corrosion. **(2)**

(d) Any two from: greasing, painting, electroplating **(2)**

2 (a) Any two from: same volume of water, same temperature, identical nails, same mass of metal, same surface area of metal **(2)**

(b) The nails would have rusted in test tubes 1 and 4. They would not have rusted in test tubes 2 and 3. This is because zinc and magnesium are more reactive than iron and would have corroded instead of the iron.

Copper and silver are less reactive than iron, so the nails would have acted sacrificially for the copper and silver, and rusting would have been faster in test tubes 1 and 4. **(3)**

(c) The results, would be the same as steel is an alloy containing iron. **(2)**

105. Alloys

1 (a) a mixture of metals, or a mixture of metals and non-metals **(1)**

(b) metallic **(1)**

(c) The alloy contains atoms of different sizes **(1)**, so the layers of atoms do not slide easily over each other. **(1)**

2 (a) (i) 18/24 × 100 = 75% **(2)**

(ii) number of carats = 24 × 50/100 = 12 carats **(2)**

(b) copper and zinc, instruments/door fittings/ water taps

copper and tin, statues/decorative objects

iron and carbon, engineering **(6)** **(2 for each row)**

106. Ceramics, polymers, composites

1 (a) Thermosoftening polymers soften and melt when heated, while thermosetting polymers do not soften or melt when heated. **(2)**

(b) thermosetting polymer, so they do not melt from the heat of the hob **(2)**

(c) Thermsoftening polymers contain long polymer chains that are not joined to each other **(1)** so, when heated, the intermolecular forces are broken and they can move over each other (1). Thermosetting polymers contain long polymer chains joined by cross links (1) that do not break on heating. (1)

2 (a) D (1)

(b) BO_3 (1)

107. The Haber process

1 (a) nitrogen from air (1), hydrogen from (one from) natural gas/methane/electrolysis of water/salt solution (1)

(b) Iron (1) is present and it acts as catalyst. (1)

2 (a) The reaction is reversible. (1)

(b) The gases are recycled. (1)

3 (a) $N_2(g) + 3H_2(g) \rightleftharpoons 2NH_3(g)$ (2)

(b) The reaction is reversible. (1)

4 (a) The rate of reaction becomes too slow to be commercially useful. (1)

(b) 450°C (1), 200 atm (1)

108. Fertilisers

1 (a) NPK fertilisers are formulations of various salts containing appropriate percentages of the elements nitrogen, phosphorus and potassium (NPK) (1). They are used to improve agricultural productivity. (1)

(b) A ammonium phosphate $(NH_4)_3 PO_4$, B ammonium sulfate $(NH_4)_2SO_4$ (2)

(c) $NH_3 + HNO_3 \rightarrow NH_4NO_3$ (2)

(d) $(14 \times 2)/(14 \times 2 + 1 \times 4 + 16 \times 3) \times 100 = 35\%$ (2)

(e) (i) Add an indicator and stop adding ammonia when there is a colour change. (1)

(ii) Note the volume needed with indicator and repeat the experiment with no indicator (1). Evaporate to half volume, cool and crystallise. (1)

2 (a) Potassium sulfate is soluble (1), but phosphate rock is insoluble. (1)

(b) calcium phosphate and calcium sulfate (2)

109. Extended response – Using resources

*Answer could include the following points.

450°C

- A higher temperature produces ammonia at a faster rate.
- A lower temperature gives a higher yield of ammonia, as the reaction is exothermic (Le Chatelier).
- It is a compromise between rate and yield.

200 atmospheres

- A higher pressure gives a higher yield of ammonia, as there are fewer molecules of product than reactants.
- Maintaining a high pressure is very expensive.
- It is a compromise between cost and yield.

Iron catalyst

- Its use increases the rate of both the forward and backward reactions.
- It has no effect on yield.

This answer is not exhaustive; other creditworthy responses will be awarded marks too.

110. Practice Paper 1

1 (a) methane (1)

(b) a mixture of a compound and two elements (1)

(c) filtration (1)

2 (a) (i) metallic (1)

(ii) covalent (1)

(b) Electrons are transferred from magnesium to chlorine (1). Each magnesium atom loses two electrons (1), one to each chlorine atom (1); this forms one Mg^{2+} (1) and two Cl^- ions. (1)

(c) shared pair of electrons between H and Cl, no additional hydrogen electrons on outer shell, three non-bonding pairs of electrons on chlorine outer shell (2)

* Answer could include the following points.

Melting point

- Melting point of magnesium chloride is high.
- It takes much energy to break the strong ionic bonds.
- Melting point of hydrogen chloride is low.
- It takes little energy to break the weak forces between the molecules.

Conduction

- When molten or dissolved the ions can move and carry charge in magnesium chloride.
- In hydrogen chloride there are no charged particles to move and carry charge.

This answer is not exhaustive; other creditworthy responses will be awarded marks too.

(e) hydrogen ion (1) and chloride ion (1)

3 (a) In the plum pudding model, the atom was thought to be a ball of positive charge with negative electrons embedded in it. In the nuclear model, the electrons orbit the nucleus at specific distances and the nucleus contains protons and neutrons. (4)

(b) (i)

Particle	Atomic number	Mass number	Number of protons	Number of neutrons	Number of electrons	Electronic structure
A	11	23	11	12	11	2,8,1
B	13	27	13	14	10	2,8
C	20	40	20	20	20	2,8,8,2
D	7	14	7	7	10	2,8

(4) (1 each correct row)

(ii) 3 shells = 3rd period, 1 electron in outer shell = Group 1 (2)

(iii) 3– (1)

(c) 1×10^{-10} m (1)

4 (a) 2 marks for plotting points correctly, 1 mark for a smooth line (3)

(b) The temperature increases so heat was given out. (1)

(c) burette (1)

(d) Use a lid on the cup with a hole, to allow the burette in (1). This prevents heat loss. (1)

Or

Use a pipette to measure the acid (1). It is accurate to one decimal place. (1)

(e) $NaOH + HCl \rightarrow NaCl + H_2O$ (2)

(f) $25 \times 0.10/1000 = 0.0025$ mol (2)

(g) It is completely ionised (1) in aqueous solution. (1)

(h) Add an indicator (1) and note the temperature added at the colour change. (1)

(i) M_r NaOH = 40. So concentration = $40.0/40 \times 1000/250 = 4$ mol/dm^3 (2)

5 (a) bubbles (1)

(b) Add excess magnesium carbonate to dilute sulfuric acid. Filter to remove excess magnesium carbonate. Heat the filtrate to evaporate some water, or heat to the point of crystallisation. Leave to cool so crystals form. (4)

(c) Wear safety glasses. (1)

(d) M_r $MgCO_3$ = 84, M_r $MgSO_4$ = 120; 2.1/84 = 0.025 mol $MgCO_3$; $0.025 \times 120 = 3.0$ g $MgSO_4$ (4)

(e) $1.8/3.0 \times 100 = 60\%$ (2)

(f) Some may be lost in filtering or during transfer between apparatus. (1)

6 (a) They are very small (10^{-9} m). (1)

(b) Their effects are still largely unknown. (1)

(c) carbon nanotube/graphite/diamond (1); hollow tube/layers/3D tetrahedron (1)

7 (a) phenolphthalein (colourless to pink)/ methyl orange (yellow to red)/litmus (blue to red) (2)

(b) In red wine, the colour change would not be visible. (1)

(c) (i) $18.90 + 19.00/2 = 18.95$ cm^3 (2)

(ii) $18.95 \times 0.100/1000 = 0.001895$ mol NaOH

2 mol NaOH : 1 mol tartaric acid

$0.001895/2 = 0.0009475$ mol tartaric acid in 25 cm^3 wine

$0.0009475 \times 1000/25.0 = 0.0379$ = 0.038 mol/dm^3 tartaric acid (5)

8 (a) Energy given out ΔH is from reactants line **down** to products line (1). Curve from reactants line to products line, with peak above both lines; activation energy is from reactants line **up** to the peak of curve. (1)

(b) $(4 \times C–H) + (2 \times 496) - (743 \times 2) - (463 \times 4) = -698$

$4CH + 992 - 3338 = -698$

CH = 412 kJ/mol (3)

(c) 0.2 = volume/24 so volume = 4.8 dm^3

Mass = $0.2 \times 44 = 8.8$ g (4)

9 (a)

*Answer could include the following points.

Improvement on Newland's table

- Mendeleev overcame some of the problems by leaving gaps for elements that he thought had not been discovered.
- In some places he changed the order based on atomic weights.

Comparison with modern table

- Mendeleev listed elements in his periodic table by atomic mass, but the modern periodic table lists by atomic number.

135

- In Mendeleev's periodic table there are no noble gases and there is no block of transition metals, but both are in the modern periodic table.
- In Mendeleev's periodic table there are spaces for undiscovered elements, but in the modern periodic table there are no spaces.
- Mendeleev's periodic table has fewer elements than the modern periodic table.

This answer is not exhaustive; other creditworthy responses will be awarded marks too.

(b) (i) RbF **(1)**

(ii) chlorine **(1)**

(c) (i) potassium hydroxide + hydrogen **(1)**

(ii) $K \rightarrow K^+ + e^-$ **(2)**

(d) (i) white solid, soluble in water **(1)**

(ii) potassium, chlorine **(1)**

(e) $Fe^{3+} + Al \rightarrow Fe + Al^{3+}$ **(2)**

116. Practice Paper 2

1 (a) It has a sharp melting point. **(1)**

(b) reverse osmosis **(1)**

(c) passing the water through filter beds and sterilising with ozone **(1)**

2 (a) (i) red flame **(1)**

(ii) white precipitate **(1)**, which remains with excess NaOH **(1)**

(iii) yellow precipitate **(1)**

(b) MgI_2 **(1)**

(c) Any two from: accurate, sensitive, rapid, useful for small samples **(2)**

(d) Lead the gas produced through a tube into limewater **(1)**; limewater turns milky if CO_2 present. **(1)**

3 (a) Bunsen burner **(1)**

(b) not pure/contains air **(1)**

(c) catalyst **(1)**

(d) Shake with bromine water **(1)**; the mixture becomes colourless/is decolourised with an alkene **(1)** or stays orange with an alkane.

(e) C_nH_{2n+2} **(1)**, C_3H_6 **(1)**

(f)

*Answer could include the following points.

- All are addition reactions:
  - C=C becomes a C–C.
- Conditions and products:
  - with hydrogen: catalyst, ethane
  - with water: steam and catalyst, ethanol
  - with chlorine: room temperature, dichloroethane.

This answer is not exhaustive; other creditworthy responses will be awarded marks too.

(g) (i) methanol **(1)**

(ii) only contains carbon and hydrogen **(1)**

(iii) B **(1)**

(iv) D **(1)**

(v) ethanol **(1)**

(vi) poly(ethene) **(1)**

4 (a) nitrogen **(1)**

(b) Algae and plants **(1)** evolved and produced the oxygen by photosynthesis **(1)**, so the percentage of oxygen gradually increased.

$6CO_2 + 6H_2O \rightarrow C_6H_{12}O_6 + 6O_2$ **(2)**

(c) (i) $2Cu + O_2 \rightarrow 2CuO$ **(2)**

(ii) The gas must be at room temperature when its volume is measured. **(1)**

(iii) All oxygen had reacted and was used up./Excess copper was present. **(1)**

(iv) volume gas used = 32 − 24 = 8 cm^3 **(1)**; percentage = 8/32 × 100 **(1)** = 25% **(1)**

(v) It's higher **(1)** because (one from):

- oxygen from the air in the test tube also reacted.
- there was more than 32 cm^3 of air, because of the air in test tube.
- air in the test tube also reacted, but was not measured initially. **(1)**

(d) methane **(1)**

(e) Sulfur is found in fossil fuels **(1)**. When they burn, sulfur dioxide is produced by oxidation of sulfur in the fuel **(1)** and enters the atmosphere. Sulfur dioxide can cause respiratory problems in humans **(1)** and causes acid rain. Acid rain damages plants and buildings. **(1)**

5 (a) $Mg + 2HCl \rightarrow H_2 + MgCl_2$ **(2)**

(b) Advantage (one from): convenient, easy, quick to use **(1)**

Disadvantage: reference to inaccurate measurement **(1)**

(c) sensible scales, using at least half the grid for the points **(1)**, all points correct **(2)**, best-fit line **(1)**

(d) 71–73 s (read from graph) **(1)**

(e) volume of gas = 13 cm^3 **(1)**; mean rate = 13/30 = 0.433 cm^3/s **(1)** = 0.4 **(1)** cm^3/s **(1)**, to one significant figure. For units, cm^3 s^{-1} also valid.

(f) correct tangent drawn **(1)**; 36/120 = 0.3 **(1)** (depending on graph) = 3 × 10^{-1} **(1)** cm^3/s **(1)**

(g) Use pipette/burette to measure out the acid, more accurate; repeat and take average, more accurate; take more frequent readings, better curve; use suitable method for reducing initial loss of gas, reduces error. Explanation must relate to reason. **(2)**

(h) line is steeper and to the left of original **(1)**; line A reaches 49 cm^3 sooner than the original **(1)**

(i) There would be a faster reaction at 40 °C; particles have more energy, so more successful collisions with the required activation energy per second. **(3)**

6 (a) total mass = 37.5 g **(1)**; % Sn = 15/37.5 × 100 **(1)** = 40% **(1)**

(b) They have atoms of different sizes in their structures. **(1)**

7 (a) COOH and NH$_2$ circled **(2)**

(b) condensation (polymerisation) **(1)**

(c) (i) so that it does not run on the paper **(1)**

(ii) three **(1)**

(iii) distance moved by X = 5.4 cm **(1)**; distance moved by solvent = 7.2 cm **(1)**; R_f value = 5.4/7.2 = 0.75 **(1)**; X = leucine **(1)**

8 (a) Catalysts change the rate of chemical reactions but are not used up during the reaction. **(2)**

(b) (i) The amount of CO$_2$ (product) will decrease **(1)**, because the forward reaction is exothermic, so the equilibrium will move to the left. **(1)**

(ii) More CO$_2$ will be produced **(1)**, because the system moves to fewer molecules (4 moles reactants to 3 moles products) under a higher pressure. **(1)**

9 (a) Phytomining uses plants to absorb copper compounds from the growing medium **(1)**. The plants are harvested and then burned to produce ash that contains the copper compounds **(1)**. Bioleaching uses bacteria **(1)** to produce leachate solutions that contain copper compounds **(1)**. Copper can be obtained from solutions of the copper compounds by displacement using scrap iron, for example, Fe + $CuSO_4 \rightarrow Cu + FeSO_4$. **(1)**

(b) Any two from: less energy needed, plants release energy when burned, less waste produced, low-grade ores can be used, no release of sulfur dioxide **(2)**

Notes

Notes

Notes

Notes

Notes

Published by Pearson Education Limited, 80 Strand, London, WC2R 0RL.

www.pearsonschoolsandfecolleges.co.uk

Text and illustrations © Pearson 2016
Typeset, illustrated and produced by Phoenix Photosetting
Cover illustration by Miriam Sturdee

The right of Nora Henry to be identified as author of this work has been asserted by her in accordance with the Copyright, Designs and Patents Act 1988.

First published 2016

19 18 17 16
10 9 8 7 6 5 4 3 2 1

British Library Cataloguing in Publication Data
A catalogue record for this book is available from the British Library

ISBN 978 1 292 13126 9

Printed in Slovakia by Neografia

Acknowledgements
Content written by Iain Brand is included in this book.

Content from the following titles is included.
Revise AQA GCSE Chemistry (9-1) Revision Workbook Higher 978129131269
Revise AQA GCSE Science Revision Workbook Higher 9781447942153
Revise AQA GCSE Additional Science Revision Workbook Higher 9781447942474
Revise AQA GCSE Extension Science Revision Workbook 9781447942160

Note from the publisher
Pearson has robust editorial processes, including answer and fact checks, to ensure the accuracy of the content in this publication, and every effort is made to ensure this publication is free of errors. We are, however, only human, and occasionally errors do occur. Pearson is not liable for any misunderstandings that arise as a result of errors in this publication, but it is our priority to ensure that the content is accurate. If you spot an error, please do contact us at resourcescorrections@pearson.com so we can make sure it is corrected.

Printed in Great Britain
by Amazon